Cleaning Techniques in Conservation Practice

Edited by
Kyle C. Normandin
Deborah Slaton

Consultant
Norman R. Weiss

Managing Editor
Jill Pearce

A Special Issue of the JOURNAL OF Architectural Conservation

Routledge
Taylor & Francis Group

LONDON AND NEW YORK

First published 2005 by Donhead Publishing Ltd

Published 2015 by Routledge
2 Park Square, Milton Park, Abingdon, Oxon OX14 4RN
711 Third Avenue, New York, NY 10017, USA

Routledge is an imprint of the Taylor & Francis Group, an informa business

ISBN 13: 978-1-873394-74-8 (pbk)

ISSN 1355-6207

A CIP catalogue for this book is available from the British Library

Cleaning Techniques in Conservation Practice is a special edition of the *Journal of Architectural Conservation* and is the November 2005 (Volume 11, Number 3) issue of a subscription.

Cover Photograph: The scaffolding to the Dome and Cone at St Paul's Cathedral designed by RDG Engineering and erected by Palmers Ltd. (Angelo Hornak)

Contents

Introduction

Nicola Ashurst

It is always good to discuss the important subject of the cleaning of traditional building materials in an informed way. There is no better way of doing this than through practical projects which have been undertaken on the basis of sound understanding of the substrate to be cleaned, the soiling to be removed and the intimate relationship between the two. This issue of the *Journal of Architectural Conservation* does just that, in a series of intriguing articles from around the world, written by experienced practitioners. The issue treats you to an in depth discussion on the cleaning of an excellent diversity of subjects – the interior of St Paul's Cathedral, the concrete at Sydney Opera House, a portion of the *Titanic*, the world's largest French Gothic Cathedral (in Manhattan) and no fewer than two Saturn V rockets. Important balance is also provided in the principles of cleaning and the public perception of soiling.

Deborah Slaton and Kyle Normandin begin with a description of the techniques currently in use in North America. In the UK, we have seen all the techniques described; some are in common use, some remain in specialist niches, while others have never really 'taken off'. BS 8221: Part 1, 'Code of Practice for the Cleaning and Repair of Traditional Materials' should be read by UK readers alongside this paper for better understanding of the local situation. Slaton and Normandin also include a well presented section on applying conservation policy to cleaning, which flags up the important difference between soiling (extra, undesirable) and patina (inherent, desirable).

The White Tower at the Tower of London is the setting for Carlota Grossi and Peter Brimblecombe's paper on public perception of soiling. Constructed of Kentish Rag, a relatively light coloured stone, members of the public were canvassed about its very noticeable, streaky soiling. For most, their first impression of the White Tower was not of dirtiness, but rather grandeur of age.

Grossi and Brimblecombe's research on the White Tower and a selection of other European cathedrals has found it quite possible for the blackened surfaces of landmark historic buildings to have a level of social acceptability. This certainly contrasts with the perceptions of developer clients who see cleaning as fundamentally important to the presentation of a refurbished or redeveloped building. Cleaning has also played a key role in the urban regeneration of many UK cities. So, there are still many exciting debates to be had about how cleaning is perceived. I expect a different debate would be had depending on whether buildings are built of naturally light or naturally dark coloured stones.

Remaining in London, the article on cleaning the interior of St Paul's Cathedral reinforces the value of researching substrates, the history of the building, the history of previous cleaning and surface treatments, and the undertaking of independently commissioned *in situ* cleaning trials, and much more. To achieve the cleaning of such a significant interior, to achieve it well, on time and on budget, reflects highly on the 'homework' and preparation phase which Martin Stancliffe, Surveyor to the Fabric of St Paul's and the Doctors De Witte, share with us in their paper. The interior was cleaned using a latex film with bespoke ingredients, achieving a good level of clean whilst overcoming many of the logistical nightmares that often plague the cleaning of interiors.

The preparation phase undertaken by Akhurst, Macdonald and Waters as part of their development of a methodology of the cleaning of the folded concrete beams at Sydney Opera House is exemplary. Based on the building's *Conservation Plan* and *Design Principles* set by designer Jørn Utzon, it also included some very clearly defined and useful scientific analysis of surface characteristics, soiling and deposits, capped with interviews with labourers, supervisors and engineers involved the Opera House's construction in the 1960s. The authors' understanding of the surfaces to be cleaned and their context, on a building which will undoubtedly one day be made a World Heritage Site, is breathtaking, but so worthwhile.

The cleaning process selected (termed a 'conservation treatment'), involved steam cleaning followed by application of a thin wash of fresh calcium bicarbonate solution, left to dry slowly, precipitating calcium carbonate into the pores of the concrete. Gentle buffing during drying using methods of the original construction changed the surfaces into an opaque glaze. This recaptured the character of the concrete beams as they appeared after removal of the original formwork; a solution as unique as the building itself.

The paper on the use of high and ultra high pressure waterjetting techniques in the conservation of historic metals comes to life immediately when you find out that its case studies are based on a portion of the *Titanic* and two Saturn V rockets. The techniques discussed will be of greatest interest to those involved in the cleaning of external ironwork and bronze-work, either sculptural or architectural. Waterjetting technology which involves increasing the power of water, was reported on in the 1960's when Dr Norman Franz, a forestry engineer and professor at the University of British Columbia, was looking for a faster method of slicing large trees into 'lumber'. The article provides a very clear explanation of waterjetting, its many adjustable parameters and how these can be fine-tuned to the metal substrates and the cleaning result required. The selectivity and sensitivity of the process, designed by the right people, on the basis of well researched investigations, is impressive. Unfortunately waterjetting cannot be as pre-dictably controllable on masonry because the characteristics of masonry and its joints are too variable. The article makes several memorable points. My favourite is: 'Dirt or unwanted material in crevices and cracks cannot be reached by a particle that is larger than the crevice opening, whereas water is not limited by the size of the pits or crevices.' We should remem-ber this when considering abrasive cleaning on masonry.

Martin Cooper's article on laser cleaning clarifies what laser cleaning is and how it operates. He explains how laser cleaning is an extremely high quality method of cleaning which can be finely tuned to remove only the soiling from the finest and most delicate of surfaces. There is also a reas-suring warning that, as with all cleaning techniques, damage to a surface will result if laser cleaning is carried out poorly.

At present, the bulk of laser cleaning undertaken in the UK takes place in museums or conservator workshops. In these contexts, the rate of clean-ing can be as fast if not faster than other comparable techniques available in the conservator's toolbox.

Laser cleaning equipment which has been modified for building site con-ditions is also available and is being used primarily for cleaning monu-ments and sculptural and architectural detail on buildings. The logistical considerations of on-site laser cleaning are very clearly explained in this paper.

Most of us still consider laser cleaning as a relatively slow specialist process which is best suited to sculpture. However, in Europe whole façades have been cleaned by laser, e.g. the entire front façade of Rotterdam's City Hall. There, the laser unit remained on the ground linked by 45 metre long optical fibres to cleaning operatives in a 'cherry picker' personnel lift. This is an excellent example of thinking outside the box which should encourage us all to stretch our minds when designing any cleaning regime.

A fire in the Cathedral Church of Saint John the Divine in Manhattan, New York City, provided the material for the article by Kavenagh and Gembinski. During the fire, thick black smoke penetrated the entire build-ing interior. The subsequent insurance claim required comprehensive and irrefutable documentation that all internal surfaces required cleaning because products of the fire had deposited on every surface. Cost control provided the mechanism for an in-depth analysis of the materials deposited on the granite and limestone interior, and the distribution of this.

Once this was established, on-site trials eliminated both chemical and abrasive cleaning processes as none tested successfully, and none had operational logistics that could fit in the environment of a busy, operational church.

Latex, with a chemical additive, was tested and proved successful on light and moderately soiled stone, but not on the heavily soiled stone. No method was found that could clean these surfaces where bitumen globules were the prime deposit.

All the projects described in the papers are fascinating to read about and learn from. Even when simple and less prominent buildings are cleaned, many of the principles and procedures described here can be applied.

It is my experience when it comes to masonry cleaning that assumptions are never wise. This is certainly true in the field of traditional masonry cleaning where the temptation must be resisted. We are particularly prone to assumptions about soiling.

On three projects over the last two years I commissioned petrographic analysis to determine the nature of the soiling, the nature and condition of the stone substrate and the inter-relationship between the two. No one was more surprised with the results than I. A Portland stone building on one of London's most trafficked 6-lane roads exhibited a moderate level of black soiling in a disfiguring weathering pattern. It was assumed that the diesel vehicle particulate emissions would be a predominant ingredient of the soiling. However, the analysis proved there was no carbon at all.

A large garden structure of sandstone located in a part of Yorkshire, heavily polluted with industrial emissions in the nineteenth century, was assumed to have a largely carbon-based soiling. Analysis proved the assumption wrong as it identified the soiling as a finely textured mat of organic growth.

Analysis of 1 cm thick black crusts on a magnesium limestone railway viaduct in an area of Nottinghamshire that was heavily industrialized in the nineteenth century, confirmed the crust to have no carbon constituents. Instead, it was found to be a thick layer of gypsum, formed by the reaction of the stone with sulphur, probably generated by the burning of coal in the area. The calcium required for the gypsum had been leached from the stonework itself, creating a highly porous layer of stone just beneath the surface. This porous layer was, in turn, protected by a thin layer of gypsum. There was no 'clean break' between the stone and the soiling.

We need to go back to first principles on every cleaning project in which we become involved. We need to check that our assumptions are correct, that we really do know about the materials we are cleaning and the soiling we are removing from them. Carefully selected analytical procedures will provide this, often at minimal cost. And we need to understand in detail the actual effect of what our selected cleaning processes are doing, before we inflict them on a whole building.

We also need to remember that there is still a lot of merit in the more common water-based, chemical-based and abrasive cleaning processes that have been in use for the last twenty years. The increased sophistication of many of these is producing excellent results, especially in the hands of skilled specialist contractors, despite the fact that they do not have glossy brochures.

The advertising of cleaning materials and equipment has become more sophisticated with the result that certain processes are better known for their name than they are for what they do. In all areas of consumer life, we need to stay vigilant, and this includes the cleaning of traditional materials.

Any preference for describing cleaning systems by proprietary name is best avoided – it helps us forget the fundamentals on which a cleaning process operates, and to lapse into a false sense of security that, at last, the cleaning process suited to every substrate, substrate condition, and to every soiling type has been found. There is a wish in all of us that one day a miracle cure will be invented which means we no longer have to understand substrates and soiling. Several brochures for cleaning systems try to persuade us we are there, at least nearly. The reality is that we still have to be informed and experienced about cleaning and, if we are not, we need to find someone who is.

Not many of us have the luxury of cleaning external façades that have not been cleaned before. The effect of previous cleaning regimes must be understood before any further cleaning is undertaken as it can significantly alter the selection of an appropriate cleaning regime. It is always worth talking to the operatives who undertook the work as they will remember what was actually done rather than what was specified.

Each cleaning operation has an effect on the masonry substrate. While the best designed cleaning regimes keep this to a minimum, the incremental effect of repeated cleaning will have a significant impact. This is of particular concern when lease terms require façade cleaning at short and regular frequency. There is no better example of this than the repeated removal of graffiti, where the incremental effect of even careful use of chemical cleaners and pressure water can soon produce noticeable damage.

We need to use more descriptive terms for cleaning. The word cleaning is not sufficient for the separation and removal of a wide variety of soiling types, from an even wider range of substrate types and substrate conditions to which they are bonded, in an equally complex range of ways.

We have come a long way in the field of masonry cleaning in the last 20 years. We also have much more to learn. The following papers are an excellent start in this education process.

Biography

Nicola Ashurst
Nicola is an established specialist in the conservation and cleaning of historic buildings. She was formerly working for the Research and Technical Advisory Services of English Heritage and is now Principal of Adriel Consultancy, Melrose, Scotland. Nicola is author of *Cleaning Historic Buildings*, Volumes one and two.

Masonry Cleaning Technologies

Overview of Current Practice and Techniques

Deborah Slaton and Kyle C. Normandin

Abstract

The goal of a masonry cleaning programme is typically to remove surface contaminants without damaging the substrate. Cleaning techniques are continually being refined in an effort to provide more effective and environmentally sustainable methods with less potential to damage substrates. Specific goals of new technologies address the preservation of patina while systematically removing various types of soiling deposits, staining, organic growth, and coatings. This paper provides a summary overview of cleaning technologies and techniques currently used in conservation practice as applicable to masonry.

For systems presently used for cleaning masonry, the paper focuses on appropriateness for different substrates and conditions based on laboratory studies, case study examples, and published literature. Systems addressed include water and steam cleaning methods; wet and dry abrasive systems; laser-based cleaning systems; chemical cleaners; coating removal systems; and biocides. The paper examines the conservation precepts and standards that provide a framework for cleaning projects and provides a proposed methodology for masonry cleaning. Issues addressed include the establishment of cleaning criteria, understanding the nature of the substrate and soiling, evaluating advantages and disadvantages of cleaning systems, developing project-specific documents, the importance of in situ trial samples, and providing quality control.

A philosophy of cleaning

Varying perspectives on whether to clean

The arguments for and against the cleaning of historic structures need to be carefully considered as a background to the evaluation of cleaning techniques currently in use. Although cleaning can be successful in removing deleterious contaminants from a building façade, cleaning does not necessarily help preserve the material and historic fabric of a building. Although in some cases cleaning is necessary to assist in the repair or maintenance of historic monuments and structures, in many cases the primary reasons for cleaning may be for aesthetic purposes or simply to reveal the underlying appearance of the building.

One of the principal arguments in favour of cleaning programmes for historic masonry is that cleaning can reduce potential soil build-up or possible damage related to contaminant deposits. Cyclical cleaning programmes may remove surface deposits of soot or contaminant build-up that, if not removed, may cause surface spalling; such deterioration in turn could necessitate more extensive material replacement in the future. However, careful examination of existing conditions is needed to determine whether the overriding reason for cleaning is to 'improve' the appearance of the historic structure, which is not necessarily an appropriate goal in conservation cleaning. In contrast, it has been argued that soiling build-up combined with interrelated patterns of surface erosion is part of the historic appearance and consequently, part of the historic structure. In some cases, it may be challenging to differentiate layers of atmospheric build-up and soiling patterns from the patina[1] that is part of the age and character defining features of the structure.

The maintenance practice of cleaning building façades at regular intervals has not always been perceived to be advantageous in conservation building practice. The *Manifesto* of the Society for the Protection of Ancient Buildings (SPAB), written by William Morris in April 1877, was strongly influenced by the concept of stewardship of historic monuments, also expressed by John Ruskin. In *The Seven Lamps of Architecture*, Ruskin noted a new interest in ancient monuments with increasing concern for the changes and potential destruction that could result from 'restoration' practices.[2] Instead, minimal intervention was proposed, to 'stave off decay by daily care, to prop a perilous wall or mend a leaky roof by such means as are obviously meant for support or covering ...'[3] As promoted by the *Manifesto*, if cleaning by gentle rubbing, dusting, or light rinsing is sufficient to remove surface soiling, then this limited treatment is an attractive conservation approach.

In France during the same time period, Emperor Napoleon III enacted a law requiring that building stone in monuments be maintained and cleaned every ten years. Nearly eighty years later, in 1958, André Malraux as Minister of State for Cultural Affairs in France was responsible for a renewal of cleaning programmes in Paris. Malraux instituted statewide programmes for cleaning the façades of the Louvre and other monuments, an effort considered by many critics to be an act of vandalism due to systematic removal from building surfaces of the patinas that reflected the historic character so familiar to and revered by the public. Contrary to the intent of such programmes for the aesthetic improvement of public monuments, the critics and general public did not necessarily perceive the overall cleaning of monuments in order to enhance their appearance. But, to Malraux, as custodian of the monuments and Minister of State for Cultural Affairs, instituting cleaning initiatives was not only to be considered as an aesthetic improvement but also a necessary part of overall maintenance and conservation.

Current guidelines

Recently developed codes of practice provide guidance on the selection of sensitive cleaning methods for the removal of deposits from the surfaces of building substrates. A guide developed by the American Society for Testing and Materials (ASTM) in North America, which is currently in the process of review and re-ratification, provides assistance in this regard.[4] The ASTM guide addresses selection and assessment of cleaning techniques for removing soiling and staining from masonry, concrete, and stucco surfaces, including a summary of cleaning techniques and commentary on testing and evaluation of cleaning samples. Interestingly, the guide notes that in some cases, cleaning may be inconsistent with the goals of historic preservation. The approach of not cleaning is recognized as a viable option.

British Standard BS 8221-1:2000 'Code of practice for cleaning and surface repair of buildings',[5] of which Part I addresses cleaning of natural stones, brick, terracotta, and concrete, came into effect on 15 April 2000. Specifically, the code of practice provides guidelines and recommendations on the selection and application of cleaning methods for external façades of buildings and is meant to be used in conjunction with BS 8221-2 which provides guidelines for surface repairs of masonry. This standard is not meant to represent a specification on how cleaning should be performed but rather, it outlines the range of practice that also includes identification of characteristics of deposits and the susceptibilities of substrates to different cleaning processes. The code of practice is intended to provide general information on cleaning in order to 'enhance a building's appearance' and/or to assist in the maintenance and/or conservation. As part of a maintenance and/or conservation programme, the goal of cleaning is described as the removal of harmful deposits from the building fabric to help prevent future decay. Consideration of cleaning is also recommended when necessary to prepare the surface for additional repair treatments, and finally, cleaning is considered when necessary to 'fulfill the terms of lease that requires periodic cleaning of a building'.

Concurrent with the ongoing development of guidelines for cleaning is a lengthy debate on whether heritage buildings should be cleaned. Over the last decade there has been increasing concern, particularly expressed in publications by Historic Scotland, that cleaning of stonework may entail harmful and damaging physical effects to the historic fabric and the overall aesthetic of buildings. In order to respond to this concern, a number of research programmes have been implemented resulting in publications including the proceedings of the international conference held Edinburgh in April 1992.[6] Other related publications include *Stonecleaning: A Guide for Practitioners*,[7] which focuses on soiling of buildings, aesthetic issues, testing methodology, physical and chemical cleaning methods, health and safety, and planning issues.

The standards and guidelines developed in recent years for cleaning of heritage structures are based on the principles of preservation, conservation, maintenance and repair, as defined by various documents that generally follow the precepts of the *Venice Charter* of 1964.[8] It is well known that conservation terminology varies from one language to another. Internationally, the word conservation has frequently been used in a broad context, primarily beginning with the *Athens Charter* of 1931,[9] where the term conservation is used to imply the safeguarding of architectural heritage. While the *Venice Charter* discusses conservation in a total of five articles (Articles 4 through 8), it does not provide or reference a specific definition of this term. Conservation is generally defined as regular maintenance and as an appropriate goal for care of a historic building. Similarly, the Australia ICOMOS *Burra Charter* of 1999[10] builds upon principles established in the *Venice Charter*; however, the term conservation is expanded to mean a multi-disciplinary activity based on a scientific approach to preserving the 'significance of place' instead of referring only to the care of a historic building. In this case, the concept of 'place' is used to broaden the meaning of a monument, building, or site. For example, Article 3 of the *Burra Charter* reads as follows:[11]

Article 3: Conservation is based on a respect for the existing fabric and should involve the least possible physical intervention. It should not distort the evidence provided by the fabric.

The responsibilities and custodial duties of conservators are ongoing and several organizations offer codes of ethics that provide guidelines on how to ensure an appropriate philosophical approach to conservation. The American Institute for the Conservation of Historic & Artistic Works (AIC) *Code of Ethics* recommends that the conservator:[12]

> ... endeavor to use only techniques and materials which to the best of current knowledge, will not endanger the cultural and physical integrity of the cultural property. Ideally, these techniques and materials should not impede future treatment or examination ... Whenever possible, the conservator shall select the techniques which have the least adverse effect on the cultural property.

Similar considerations are addressed by the International Institute for Conservation of Historic and Artistic Works (IIC) Canadian Group *Code of Ethics and Guidance for Practice for those involved in the Conservation of Cultural Property in Canada*, written in Ottawa in 1986,[13] as well as by publications of various other conservation organizations.

The US Secretary of the Interior's Standards for the Treatment of Historic Properties provide similar guidance for approaching the care of historic structures. Also, specific to cleaning, the Standards state: 'Chemical or physical treatments, if appropriate, will be undertaken using the gentlest means possible. Treatments that cause damage to historic materials will not be used.'[14] Among the guidelines that accompany the Secretary of the Interior's Standards for Preservation are the following recommendations for exterior masonry:[15]

> Clean masonry only when necessary to halt deterioration or remove heavy soiling.
>
> Carry out masonry surface cleaning tests after it has been determined that such cleaning is appropriate. Tests should be observed over a sufficient period of time so that both the immediate and the long range effects are known to enable selection of the gentlest method possible.
>
> Clean masonry surfaces with the gentlest method possible, such as low pressure water and detergents, using natural bristle brushes.

Careful consideration of whether cleaning is necessary requires that numerous questions be answered, including the following. Why does the masonry need to be cleaned? Will the masonry deteriorate if it is (or is not) cleaned? Can the building be cleaned without causing damage? Even if these questions are addressed, and detailed attention is given to the proper selection and application of appropriate methods and techniques to be used for cleaning, another important question still remains: How clean is clean enough? Certainly, much depends on the material substrate to be cleaned. Each cultural monument, building, or historic structure must be considered on a case-by-case basis in consultation with professionals who are knowledgeable in heritage conservation and practice. It should be noted that local governments can also provide ordinances with cautionary provisions that require review and monitoring of the cleaning process.

Goals and approaches to cleaning

A successful approach to any masonry cleaning project must consider the goals and context of the cleaning process. The conservation precepts outlined in the standards mentioned above provide an appropriate framework for any cleaning project, whether or not the building or structure is architecturally or historically significant. A proposed methodology for masonry cleaning involves the following:

1 establish criteria for the cleaning project
2 understand the nature and condition of the masonry substrate and the soiling to be removed
3 evaluate the advantages and disadvantages of possible choices of cleaning systems
4 develop appropriate contract documents (method statement, drawings, and specifications) for site-specific cleaning
5 provide quality control during building cleaning

Establishing criteria for a cleaning project requires taking into account the goals or reasons for cleaning. As discussed above, cleaning may be performed to remove deleterious contaminants or failed coatings, to prepare a substrate for further treatment or repairs, or simply to improve the appearance of the building or structure. As part of establishing these criteria, the philosophical issue must be addressed of whether accumulated soiling represents a character defining patina, and as such, should be preserved.

Understanding the physical and chemical nature and condition of the soiling to be removed as well as of the building materials to be cleaned is critical to successful cleaning. Soiling may consist primarily of atmospheric pollutants, other contaminants, biological growth, staining related to components of the masonry substrate, or a combination of these factors. For example, on buildings in urban centres with heavy vehicular traffic, accumulation of carbon deposits related to buses and cars is common. On urban buildings that have remained uncleaned over an extended period of time, the use of oils and coal for heating often results in similar deposits. In moist climates, soiling may be more typically related to biological growth, particularly on shaded north facing elevations and horizontal façade elements. On some masonry substrates, localized staining may be related to exposure of ferrous minerals in the substrate to acids. In each case, the composition of the soiling needs to be determined, to evaluate whether the soiling is deleterious to the substrate and also to define whether the soiling is part of natural patination of the masonry through weathering processes.

Understanding the condition of the masonry is also critical to successful cleaning. For example, a fragile surface is generally not a good candidate for cleaning with systems involving water or abrasives applied under moderate or high pressures. As another example, stone containing ferrous mineral inclusions may be subject to staining if cleaned with strong acidic chemicals. Certain light coloured stones are vulnerable to staining if cleaned with water having high iron content, even without the use of chemical cleaners. An existing coating or well-established organic growth accumulation may be resistant to removal with abrasive methods. In all cases, it must be remembered that if the substrate cannot be successfully cleaned without damage, then less cleaning or no cleaning at all is always a possible alternative.

Evaluation of the possible choices of masonry cleaning systems is becoming increasingly complex as more types of cleaning systems – and systems of greater complexity appear on the market. Improper cleaning can damage the substrate by causing damage such as staining, discoloration, etching, or erosion. Over-cleaning, by which is meant either an overly aggressive single cleaning programme or too frequent repetition of cleaning, can result in deterioration of the substrate. Such damage can increase the likelihood and rate of future dirt accumulation, absorption of moisture, and surface deterioration. In addition, some cleaning systems can present a potential source of damage to other building elements and materials, nearby structures, and the environment, as well as a hazard to workers, other persons, and animals in the work area. The process of evaluation to select the appropriate cleaning system typically involves research, laboratory and *in situ* trials, and testing of multiple systems before one or more is selected for use on large-scale *in situ* samples, followed by overall implementation.

Once the appropriate system is selected, contract documents are developed using the information gathered through the evaluation process. Proper drawings and detailed specifications, based on a thorough examination of the historic structure and evaluation of the proposed cleaning system, are critical to understanding the extent of the work to be undertaken, as well as how the work will be performed. Sample test areas carried out at the beginning of work are used to establish a standard and as reference for the remainder of the cleaning work. Finally, involvement by the professional during cleaning, either in performing the work or as a reviewer and observer of the work in progress, ensures continuing quality control during the cleaning.

Overview of masonry cleaning techniques

The following overview highlights the primary systems available today, concentrating on applications for exterior surfaces composed of stone, brick, terracotta, concrete, and stucco. Masonry cleaning techniques available today include traditional methods (e.g. water and steam) as well as new techniques that have come into use in the past two decades (e.g. microabrasive systems). In addition, systems initially designed for use in other industries (e.g. pelletized carbon dioxide cleaning), or for use with objects and sculpture, (e.g. laser cleaning) are increasingly being applied to the cleaning of building façades.[16]

Current cleaning systems used on masonry buildings can be characterized in four categories: water, abrasive, laser, and chemical techniques. Depending on the nature and condition of the substrate, and on the character of the soiling to be removed, systems or a combination of systems may be appropriate for a specific cleaning campaign. Other criteria to be considered include protection requirements, environmental constraints, special application requirements, and cost.

In addition to the stated goals of the cleaning programme, other factors in selection of a cleaning system include the nature and condition of the substrate, contaminants to be removed, site and temperature constraints, protection and disposal requirements, and cost. Also, the level of cleanliness desired may be a factor in identifying the cleaning systems that can be used in a particular application. For example, if some remaining soiling produces an acceptable cleaning result, then it may be possible to use a gentler cleaning system than would be needed if more thorough removal of soiling is required. Variables in cleaning systems include formulations and dilutions of cleaning chemicals, materials used in abrasive techniques, and application pressures for cleaning materials and for water used in prewetting, cleaning, and rinsing, among other factors, which are further discussed below.

Water and steam systems

In water cleaning techniques, loosely bonded atmospheric dust and dirt, soot, and other deposits can be softened or dissolved by brushing, soaking, or mist spraying, and the surface then rinsed to remove the residual dirt. An important factor in the use of water systems is the pressure at which the water is applied to the wall. This consideration must also be given to chemical systems, many of which involve application of water for prewetting and rinsing. Depending on the substrate, the definition of low, medium, and high pressures varies widely; an application water pressure that is considered high for masonry may in turn be low for use on some metals.

Figure 1 For low pressure water soak and rinse cleaning, spray racks with multiple nozzles are used. The spray rack can be configured to reach complex and difficult profiles on the building façade. This technique has been found to be very effective on limestone and marble substrates.

Gentle water cleaning methods based upon intermittent misting or continuous soaking use very low pressures (less than 50 psi), followed by a low (100 to 200 psi) or medium (200 to 400 psi) pressure water rinse. Moderate to high pressure water washing (400 to 800 psi) is generally safe for use on stone such as granite.[17] However, water washing at these higher pressures can potentially erode soft or fragile stone surfaces.[18] The effect of the water spray or stream on the substrate is also related to the volume of water, the aperture of the nozzle, and the distance of the nozzle from the wall surface. For example, low pressure water cleaning techniques often involve the use of multiple fan-spray nozzles attached to a spray rack suspended adjacent to the wall surface being cleaned. Specialized rack systems can be built to clean complex surface profiles, and low pressure sprays may be continuous or intermittent. In contrast, high pressure water cleaning is typically performed from a single nozzle with a very narrow fan spray. A narrow aperture or a lower water volume at a given application pressure will generally result in higher pressure as water reaches the wall. Low pressure water misting systems are typically most successful on calcareous materials such as limestone and marble, as atmospheric soiling is readily loosened from these surfaces by gentle methods. Also, water (or microabrasive) systems are preferred to chemical cleaners for use on

Figure 2 For very fine mist cleaning, adjustable nozzles are conformed to reach many areas of delicate stone crevices. Each nozzle is equipped with an arm that can be adjusted to reach difficult areas of access where soot and carbon build-up can occur.

marble and limestone because siliceous components of the stone are dissolved by certain acids.

Successful water cleaning depends on proper selection of application techniques and pressures, proper protection methods prior to cleaning, and skill of the operative in maintaining a controlled application. In addition to issues related to water pressure, problems associated with improper water

Figure 3 A close-up view of the system shown in Figure 2.

cleaning include water infiltration through open joints or cracks, or through the face of porous masonry materials. If these concerns exist, dry cleaning systems may need to be considered. Care must be taken to avoid excessive water infiltration or deep saturation of masonry substrates. For example, damage can occur due to cyclic freezing and thawing of entrapped moisture if water cleaning is performed in cold climates during winter weather.

Steam cleaning offers the advantage of using less water than most water cleaning techniques, and is therefore a good alternative if the volume of water used in the cleaning process needs to be limited because of conditions of the substrate or site constraints. Steam cleaning was not used for some time because of hazards to workers, but has become available during the past decade through improved equipment that provides better control and safety.[19]

Any water used in cleaning should be checked to confirm that it is free from impurities or minerals that could stain the substrate. For example, only a small amount of iron present in the water source used for cleaning can result in staining of a sensitive substrate such as white marble.

Figure 4 The trained operator is using a proprietary high temperature steam system. This system utilizes low pressures of steam to clean the substrate. The combined low pressure and heat of the steam can be used alone, or as a rinse procedure to improve results obtained with other cleaning procedures.

Abrasive systems

Abrasive systems clean by abrading, roughening, or eroding the surface; however, the extent to which this abrasion is considered as damage to the substrate depends upon the nature of the substrate and the aggressiveness of the system used. Mechanical cleaning methods using high-pressure sand or grit blasting are no longer in common use for historic masonry because of damage caused to the substrate by such cleaning. In contrast to this type of abrasive technique, new microabrasive systems have come into use during the past few decades that clean by directing fine particles at the wall at very low pressures (i.e. typically between 25 psi and 75 psi).[20] Unlike older methods, these techniques when used with proper controls can remove contaminants from delicate surfaces without significantly damaging the material. Abrasive and microabrasive systems vary in the type of materials and specialized nozzles or other equipment used in application, and whether the system involves water for cleaning and rinsing. Most microabrasive systems have in common the use of very fine particulates for cleaning applied at very low pressures. As these systems depend on compressed air, other variables that affect the cleaning process include air pressure and flow rate.

Examples of microabrasive systems presently available include a wet[21] and a dry[22] proprietary low pressure cleaning system using specialized nozzles.[23] The cleaning process can be adjusted by varying the material, hardness, size, shape, and type of particulate); varying the nozzle size, and application pressure. Some proprietary dry microabrasive systems include a containment system for the particulate and debris removed from the building.[24] All of these techniques use very small particulate (less than 90 microns in diameter), applied with compressed air at very low pressures (typically 25 to 75 psi maximum).

Figure 5 Microabrasive system (Sponge-Jet). The trained operator is using a proprietary microabrasive system with fine sponge media and embedded soft dolomitic particles.

Figure 6 Microabrasive system (Façade Gommage®). The trained operator is using a proprietary microabrasive system with fine mixed media comprised of soft silica, alumina, and soft organic particles.

Figure 7 Microabrasive system (Rotec Vortex). The trained operator is using a proprietary microabrasive system with fine silica based media.

Another proprietary system uses open cell water-based polyurethane sponge to which various types of abrasives of varying sizes, ranging from hard alumina to soft dolomitic particles, are bonded.[25] When applied to the surface at very low pressures (typically 35 to 60 psi), the sponge flattens out and the abrasives remove surface dirt layers from the substrate. Contaminants are contained by the sponge after impact, reducing airborne dust. The sponge particles reshape after impact and can be recycled for reuse. This system can be used as a microabrasive cleaning technique or at higher pressures and with different grit to provide a more forceful cleaning process where needed in special applications.[26]

Sodium bicarbonate (baking soda) is another medium used in microabrasive cleaning. This material has been found to be problematic in some applications, since baking soda and water form a paste that is difficult to remove. In addition, sodium bicarbonate is a salt and if not thoroughly removed by rinsing can be damaging to masonry. Another system uses compressed air to accelerate frozen carbon dioxide (CO_2) or 'dry ice' pellets to a high velocity. When the pellet strikes the substrate, the rapid transfer of kinetic energy and heat between the pellet and the substrate cause the frozen pellet to shatter and vaporize, removing coatings or contaminants from the substrate surface. This technique has been used with success on metal equipment but does not appear to have had widespread success in cleaning masonry.

Even microabrasive systems can damage historic masonry and should only be implemented after a thorough review including laboratory and *in situ* trials and evaluation. Microabrasive systems have typically been used with success to clean building stone such as sandstone and granite.

Laser cleaning

Laser-based cleaning systems (initially used for cleaning monuments and artwork) have gained increased recognition over the last two decades for their successful use in building cleaning. These systems operate by application of laser in a pulse frequency in which the power output is controlled. The handheld devices can focus the laser beam to a fine point, thereby adjusting the intensity of the laser impact over the surface. Given the fineness of the laser point, it is often time consuming to achieve an even appearance with laser cleaning, particularly on large surface areas. Although application to objects and monuments has shown promise for the effectiveness of laser cleaning, its use in large scale applications has to date been very limited.[27]

Chemical systems

In chemical cleaning systems, the products react with the dirt layers or the substrate to dissolve or remove soiling and stains. Because many chemical cleaners are used with water, the same considerations for appropriate water pressures and the need to prevent leakage through open joints in the building apply. In addition, and of great concern, many chemical cleaners contain strong alkalis or acids that are very hazardous to humans and animals, and can also cause damage to the materials being cleaned as well as to other building elements, site features, plants, the environment, and even adjacent vehicles and buildings. Special collection and disposal procedures may be required to control run off from chemical cleaning.

Chemical cleaning compounds can be alkaline or acidic cleaners, or organic solvents, provided in liquid, gel, or poultice form. In addition to the composition and character of the cleaning compound, other factors in their application include dilution, application technique, and dwell time.[28] While product literature generally provides a description of the cleaning product and information on how to use it, it is necessary to obtain detailed information about the product's composition to understand fully the specific safety and protection issues.[29] However, even material safety data sheets may not provide complete information about the cleaner since these documents typically only report what is required by law.

Two alkaline compounds commonly used in masonry cleaning are trisodium phosphate (TSP) and sodium hydroxide solution. These cleaners may be used alone or as a pre-wash in combination with a mild acid after-wash. In this case, the alkaline pre-wash reacts with the soiling, and the acidic after-wash neutralizes the pre-wash. Other cleaning products contain acids. Some acids may remove soiling from some substrates but cause damage to others. Examples of acids contained in some proprietary cleaners that can cause damage to many substrates include hydrofluoric acid, which dissolves silica and other siliceous components such as found in many types of stone; ammonium bifluoride or sodium bifluoride, which can etch glass;[30] and hydrochloric (also known as muriatic) acid. Chemical cleaners require very careful selection and application, as they may be very hazardous to humans and animals regardless of their environmental effect or chemical description. Certain acids present in some proprietary cleaners, including hydrochloric and hydrofluoric acid, are of special concern in that they are not only hazardous but can significantly damage building materials. Strong chemicals can etch masonry and adjacent façade components, dissolve components of stone or brick masonry or concrete, mobilize severe staining, remove painted finishes, and cause other damage.

Some milder acids, particularly organic acids, tend to be potentially less harmful to users and to the substrate. Proprietary cleaners containing phosphoric, sulfamic acid, gluconic, acetic, or citric acid are generally less harmful to substrates and users. However, even milder acids may react with minerals in some masonry substrates and cause staining.

Although performance of *in situ* samples is an important step in the evaluation of any cleaning system, small scale *in situ* samples are particularly important as an initial step with chemical cleaning products. Testing of such samples should always be performed to confirm that the chemicals in the cleaner will not react with the substrate and cause staining or other damage. The surface should be pre-wet, trial samples implemented, and the sample allowed to dry after cleaning and rinsing. After cleaning the sample area should be re-evaluated following a period of at least several weeks, to confirm that staining or other reactions have not occurred.

Many chemical cleaning products also contain a surface active agent (surfactant), which acts as detergent, wetting agent, and emulsifier. Other cleaning compounds include a chelating agent such as ethylenediaminetetraacetate (EDTA), which is capable of holding metal ions. Certain chemical cleaners are particularly effective on specific stains. For example, cleaners containing oxalic acids are effective in removing some types of ferrous staining. In addition, chemical cleaners in poultices have been found to be particularly successful on deep seated or difficult to

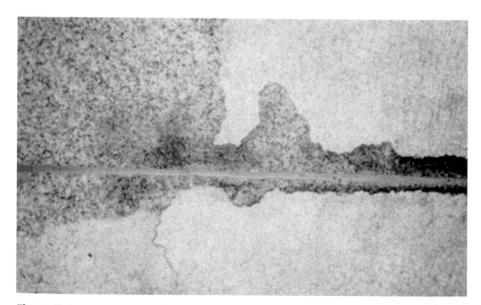

Figure 8 Ferrous staining of granite and other stones can occur if strong acids are present in cleaning products. In this case, prior cleaning with a strong acidic cleaner resulted in orange staining when the acid reacted with mineral inclusions in the granite and with iron present in the Portland cement mortar joints between the stone units. The stains reappeared each time the granite became wet.

remove stains including fire soiling, and on ornate and detailed surfaces. A poultice is a combination of a solvent and an inert absorbent material such as attapulgite, an expansive clay. Poultices are typically applied to the surface and removed after a pre-determined dwell time. Recent advances in dispersion of controlled chemical cleaning through rubber poultices are being used for interior applications, where water and microabrasive systems would be costly and difficult to use because of special protection and containment requirements.[31]

Although some chemical cleaners tend to work in removing severe stains or deposits on which other systems are not effective, these products also often have the potential to cause more extensive damage or more significant long-term effects than other systems. Therefore, these systems should not be used without professional consultation. Chemical cleaners have been successfully used on a wide variety of masonry substrates including brick, terracotta, and stone.

Coating removal

Many of the methods discussed above can also be used for coating removal; however, not all of these systems are universally successful. For example, if the bond of the existing coating is poor or failed, water techniques at pressures low enough to avoid damage to masonry substrates may remove the coating consistently from the surface. However, if the coating is well bonded, water methods may not remove any of the coating, or may only result in partial removal. Steam will sometimes remove an existing coating where water will not, but the heat of the steam can also partially dissolve some coatings, sometimes resulting in a difficult clean-up process after coating removal. The particulates used in certain microabrasive systems simply 'bounce off' some types of applied coatings without removing them from masonry, although microabrasives have been used successfully in removing coatings from metals.

In general, where coatings are intact and well adhered to the surface, chemical coating removal systems are most effective in removing them from masonry. Most paint strippers available today are either alkali or solvent based. The alkali-based products typically contain calcium hydroxide, sodium hydroxide, or potassium hydroxide. The solvent based products may contain a range of components such as N-Methyl Pyrrolidone (NMP), methylene chloride, mineral spirits, alcohols, and many other chemicals.[32] Each of these chemical products is more effective on some coatings than others and each has different environmental and safety concerns; some of these components present a significant safety hazard for users. In all cases precautions for the use of coating removal products are similar to those for chemical cleaners.

Some coating removal systems are supplied with a fibrous laminated paper mask. The stripping product, usually in gel or paste form, is applied to the substrate and covered with the paper. The coating remover is allowed to dwell for a period time,[33] after which the paper is removed together with the remnants of cleaner and debris. Some proprietary systems also come with special bags for disposal of the paper, cleaner, and debris.

Biocides

Biological growth on buildings ranges from plant growth such as ivy to microbiological growth such as lichens, algae, fungi, and bacteria. Organic growth can appear green, brown, or black, and is sometimes mistaken for accumulations of airborne dirt. Such growth is usually aesthetically undesirable and in some cases may increase the rate of deterioration of the substrate by preventing the evaporation of water. It is important to understand whether the observed stain or soiling is dirt or organic growth; cleaning techniques that remove soiling will often also remove organic growth, but that growth is likely to recur unless the cleaner incorporates a biocide.

Historically, biocides included copper and fluorides left in place on the masonry surface to inhibit the organic growth. Indeed, where runoff staining has occurred from copper flashings in stone construction, it is often noted that organic growth does not occur in the stained area. However, these types of biocides have been little used in recent years due to health concerns, and the present approach is toward cyclical application of a cleaner containing a biocide to remove and inhibit new growth.

Figure 9 Biological growth on building façades can appear green, brown, or black, and is sometimes mistaken for atmospheric pollutants and dirt deposits. The trial samples shown here included a mixture of bleach, detergent, and water, and also a proprietary biocide. Both systems were evaluated for immediate and long-term effectiveness in removing and inhibiting the recurrence of biological growth on the surface.

Figure 10a As seen in this example, biological growth has occurred on the architectural features and sculpture areas of a building. Algaecides can prove helpful in removing and inhibiting the recurrence of surface biological growth.

Figure 10b When viewed close-up, biological growth can be seen growing in crevices of ornate architectural stone carvings.

Some types of organic growth respond to biocides composed of a solution of detergent, sodium hypochlorite (household bleach), and a trisodium phosphate-type cleaner, applied to a pre-wet surface which is then scrubbed and thoroughly rinsed with low pressure water. In some cases, where it is not desirable to introduce sodium into the wall surface, calcium hypochlorite has been used in place of sodium hypochlorite. A variety of proprietary biocide products are also available to remove and inhibit organic growth. Some proprietary biocides contain quaternary ammonium chlorides,[34] which can be effective in inhibiting recurrence of organic growth. Biocides vary in their short and long term effectiveness, and only through long term trials can their success be evaluated for a given substrate. It should be noted that as these treatments are chemical in nature, small scale trials are necessary to confirm that adverse affects such as staining do not occur, prior to implementation in larger trials.

The removal of organic growth should be considered as a cyclical maintenance issue. Over time, organic growth will tend to reappear, particularly on porous substrates, in shaded areas, and in areas that remain moist over time. Reapplication of the biocide at regular intervals may be necessary to prevent reappearance of biological growth.

Masonry cleaning and quality control

To select a cleaning system, cleaning equipment, product literature and manufacturer's safety information should be reviewed to understand the nature of the cleaning products being considered. Laboratory testing of soiling and contaminants and samples of the masonry may be needed to determine what systems will be effective. Protection requirements for workers and others on-site, and for building and site features, need to be reviewed, as well as application instructions and controls.

Testing must be performed as part of the cleaning system selection and at the beginning of the cleaning project. *In situ* trial cleaning samples of each system under consideration are helpful in selecting and confirming the proper system to be used, allowing the project team to evaluate the effectiveness of the system, confirm that the proposed protection and disposal systems are adequate, and refine the cleaning procedures to be used. The level of cleanliness to be achieved also needs to be evaluated and approved by the owner and conservation professional during the sample process. In addition, the samples should be protected during the work and used as a reference for the standard of quality required. Proper project drawings and specifications document the techniques and control measures required. With any cleaning technique, the success of the application is dependent on quality control during application. Finally, field observation by the conservation professional is essential during the work to monitor the effectiveness and gentleness of the cleaning process.

Another important issue in planning for masonry cleaning is to consider the effects of repeated cleaning on the substrate. Where cleaning is performed as cyclical maintenance on a repeated basis, the effects of the processes used are cumulative. For example, if a single cleaning application results in minor erosion of the substrate, repeated cleaning by the same process over time can eventually cause noticeable loss of surface material. Such considerations support the argument that cleaning should be performed only at very long intervals, even if noticeable accumulation of soiling occurs between cleaning programmes. Sir Bernard Feilden has been quoted as saying, 'Regular inspections and maintenance, please!'[35] However, although implementation of regular cyclical maintenance of historic properties is considered good conservation practice, implementation of periodic cleaning as part of regular maintenance demands specific determination of how often the building should be cleaned.

Guidelines for masonry cleaning

The age and significance of a building or structure are understood through its historic character as depicted in its materials and features. As custodians of cultural heritage, it is critical that conservation professionals treat the fabric of historic structures and monuments with the utmost care. As noted in numerous guidelines and statements of conservation philosophy over time, a thoughtful and well-considered approach to the repair and maintenance of the historic buildings and structures, including any preventive treatments, is essential and valuable to the overall process of conservation.

It is important to note that cleaning of historic masonry is not always advisable, as the cleaning process can have the potential to cause damage to the historic fabric of the building or monument. Cleaning should be undertaken only when dirt or other contaminants may be causing damage to the masonry or obscuring significant architectural features of the structure. As part of an overall conservation programme, cleaning should not remove the patina of a building, as the patina is also part of the historic character of the structure. Cleaning methods should be carefully selected based on cleaning trials and samples that can help to evaluate how to remove soiling using the gentlest means possible without harming the fabric of the building.

The evaluation of cleaning samples and the selection of cleaning systems is typically accomplished with reliance on visual examination of laboratory and *in situ* cleaning tests, sometimes enhanced by petrographic microscopy to evaluate the effectiveness of the techniques in removing soiling and in avoiding damage to the substrate. It is often necessary to perform more than one series of tests to refine and evaluate proposed cleaning procedures. Further examination of cleaning samples may help to identify the layers of the substrate and provide helpful guidelines for defining the appropriate level and degree of cleaning. For example, initial studies indicate that the measurement of substrate characteristics during cleaning studies is a pragmatic and useful tool for evaluating the effectiveness of the cleaning system in removing contaminants and the extent to which the substrate is damaged or otherwise affected by the cleaning process.[36]

In order to define and apply cleaning criteria properly and to ensure sound conservation practice, valuable codes of ethics have been issued by numerous conservation institutions involved in the care and curatorial maintenance of our cultural heritage. Over the last two decades, technical guidelines for cleaning projects have been developed by ASTM, BSI, and others help to define criteria for the specification and application of masonry cleaning systems, including the use of laboratory analysis, *in situ* testing, and evaluation of on-site samples. These guidelines contribute to sound practice that further enhances the development of project-specific cleaning procedures. Perhaps the next step in conservation cleaning is to work toward development of criteria that define acceptable levels of cleanliness. Such criteria will assist project participants in evaluating the extent and degree of cleaning required on a project by project basis and in answering the next question to be addressed: 'How clean is clean enough?'

Biography

Deborah Slaton
Deborah Slaton is a Senior Consultant with Wiss, Janney, Elstner Associates, Inc. in Northbrook, Illinois, USA. She is a Fellow of the Association for Preservation Technology International and Vice President of the Historic Preservation Education Foundation. She has published extensively on preservation technology and is the author of a monthly column on material and construction failures for the *Construction Specifier*. Ms Slaton is co-editor of several conference proceedings including the *Preserving the Recent Past* series, and is author of the US National Park Service Preservation Brief on Historic Structure Reports.

Kyle C. Normandin
Kyle C. Normandin is a Senior Associate with Wiss, Janney, Elstner Associates, Inc. in New York City, New York, USA. He is currently a member of ICOMOS/US and is also a member of American Standards for Testing of Materials Committee E06 on Performance of Buildings. He serves as a ICOMOS US board liaison to the Association for Preservation Technology International. Mr Normandin is a member of the

International Scientific Committee on Technology for DOCOMOMO International. He is the author of numerous technical papers on various topics related to architectural conservation and preservation of heritage resources.

Acknowledgements

The authors appreciate the contributions of Joshua Freedland, Architectural Conservator with Wiss, Janney, Elstner Associates in Chicago; and Anne Grimmer, Architectural Historian, of the US National Park Service in Washington, DC to this paper.

Notes

1   The term 'patina' refers to the character of the material surface that develops with age. Patina interlayers in masonry are the thin surface layers resulting from weathering and treatment of the substrate over time. In general, patina in masonry is often regarded as a natural and positive characteristic, similar to the green corrosion layer that appears on copper and bronze and is valued for its colour and historical character.

2   Ruskin, John, *The Seven Lamps of Architecture*, Smith, Elder, and Co., London (1849), reprinted Noonday Press, New York (1961).

3   Morris, William, *Manifesto* of the Society for the Protection of Ancient Buildings (1877), www.spab.org.uk.

4   ASTM E 1857, 'Standard Guide for Selection of Cleaning Techniques for Masonry, Concrete, and Stucco Surfaces'. This guide outlines procedures for the selection and assessment of cleaning techniques for removing soiling and staining from masonry, concrete and stucco surfaces. The current version of the ASTM standard references British Standard BS6270 (1982), further discussed below.

5   The referenced standard is British Standard BS 8221-1:2000, 'Code of practice for cleaning and surface repair of buildings, Part 1: Cleaning of natural stones, brick, terracotta and concrete', issued by the British Standards Institution (BSI). Part 2 of this standard is subtitled, 'Surface repair of natural stones, brick and terracotta'. This code of practice was prepared and adopted by Technical Committee B/209 of the British Standards Institution (BSI) and organizations such as English Heritage, Historic Scotland, Society for Protection of Ancient Buildings, the Royal Institute of British Architects, and other representative councils, associations, and research institutions throughout the UK. The current standard was updated from the previous British standard adopted in 1982.

6   Webster, R. G. M. (ed.), *Stone Cleaning and the Nature, Soiling and Decay Mechanisms of Stone*, Donhead Publishing, London (1992), Proceedings of the International Conference held in Edinburgh, Scotland, 14–16 April 1992.

7   Andrew, C. A., Young M. E. and Tonge, K. H., *Stonecleaning: A Guide for Practitioners*, Historic Scotland and The Robert Gordon University, Edinburgh (1994). Also of interest is the related publication by Urquhart, D. C. M., Young, M. E. and Cameron, S. *Technical Advice Note 9, Stonecleaning of Granite Buildings*, Historic Scotland, Edinburgh (1997).

8   International Charter for the Conservation and Restoration of Monuments and Sites (*Venice Charter*), adopted by Second International Congress of Architects and Technicians of Historic Monuments, Venice, Italy, 1964.

9   *Athens Charter* adopted at the First International Congress of Architects and Technicians of Historic Monuments, Athens, Greece, 1931.

10   Australia ICOMOS *Burra Charter*, 1999.

11   Ibid.

12   *The American Institute for the Conservation of Historic & Artistic Works, Code of Ethics*, American Institute for Conservation of Historic & Artistic Works, revised 1994.

13   *Code of Ethics and Guidance for Practice for those involved in the Conservation of Cultural Property*, International Institute for Conservation of Historic and Artistic Works (IIC) Canadian Group, Canada (1986).

14   Weeks, K. D. and Grimmer, A. E., *The Secretary of the Interior's Standards for the Treatment of Historic Properties with Guidelines for Preserving, Rehabilitating, Restoring & Reconstructing Historic Buildings*, US Department of the Interior, National Park Service, Cultural Resource Stewardship and Partnerships, Heritage Preservation Services, Washington DC (1995).

15   Ibid.

16   Several references provide an overview of available cleaning techniques from the UK or North American perspective; for example, see Ashurst, N., *Cleaning Historic Buildings: Volume 1, Substrates, Soiling and Investigations; Volume 2, Cleaning Materials and Processes*, Donhead, London (1994). See also Weaver, M. E., *Conserving Buildings: A Manual of Techniques and Materials*, Wiley, New York (1997). See also, Grimmer, A., *Keeping it Clean: Removing Exterior Dirt, Paint, Stains and Graffiti from Historic Masonry Buildings*, National Park Service, US Department of the Interior, Washington DC (1988).

17   Slaton, D., 'Under Pressure', in *Construction Specifier*, December 2000.

18   In comparison, pressures of well over 30,000 psi are considered safe for use in cleaning some metal elements on buildings and other structures. Water pressures of greater than 30,000 psi have been successfully used to remove existing coatings from sound, high strength concrete substrates without damage to the concrete.

19   An example of a steam cleaning system is the Rotec steam system, distributed by Quintek of Niagara, Ontario, Canada.

20   In the United States, the term 'abrasive' is often used in reference to traditional methods such as high pressure sand blasting, which the term 'microabrasive' is typically used in reference to newer, low pressure systems using fine particulates. The negative connotation of the term 'abrasive cleaning' in the United States may be in part related to the use of high pressure sand and grit blasting on masonry structures through the 1960s, which resulted in visible damage to the walls of many buildings evaluated years after this cleaning was performed.

21   An example of a wet microabrasive masonry cleaning system is the Rotec Vortex system, distributed by Quintek of Niagara, Ontario, Canada.

22   An example of a dry microabrasive masonry cleaning system is the Façade Gommage® system, distributed by Thomann-Hanry, Inc. of Paris, France.

23   The system described is Sponge-Jet, distributed by Sponge-Jet, Inc. of Portsmouth, New Hampshire, United States.

24   As an example of the appropriate use of more aggressive sponge cleaning techniques, this system has been used by the authors to remove an alteration crust from

a granite substrate. The alteration layer was formed by a previous inappropriate cleaning of the building with strong acid cleaners, and having formed, was retaining moisture against the sound stone beneath. Thus, a more aggressive cleaning method was needed than would have been required had the alteration layer not existed.

25 Another low pressure microabrasive system, which uses a small amount of water, is the JOS system distributed by Stonehealth of Marlborough, Wiltshire, England.

26 As an example of the appropriate use of more aggressive sponge cleaning techniques, this system has been used by the authors to remove an alteration crust from a granite substrate. The alteration layer was formed by a previous inappropriate cleaning of the building with strong acid cleaners, and having formed, was retaining moisture against the sound stone beneath. Thus, a more aggressive cleaning method was needed than would have been required had the alteration layer not existed.

27 Further discussion of laser cleaning techniques is provided in publications from LACONA, including *Proceedings of the Fifth International Conference on Lasers in the Conservation of Artworks* (Osnabrueck, September 2003) Springer, Berlin (2005). See also Cooper, M., 'Recent Developments in Laser Cleaning', in *The Building Conservation Directory 1997, Cathedral Communication, Tisbury (1997)*. See also Research Report: Laser Stone Cleaning in Scotland, Historic Scotland, Edinburgh (2005).

28 Dwell time is the period for which the cleaning compound is allowed to remain on the surface. Manufacturer's literature typically provides the recommended dwell time for chemical cleaners, which can vary from les than five minutes to more than twenty-four hours.

29 Material safety data sheet (MSDS), Control of Substances Hazardous to Health (COSHH), and other health and safety (H&S) data.

30 In the presence of water, these chemicals form hydrofluoric acid.

31 The Arte Mundit® poultice, manufactured by FTB Restoration of Grobbendonk, Belgium, is an example of a system using a chemical cleaner in a latex rubber poultice. Other examples of poultice products combine cleaning compounds such as sodium bicarbonate, EDTA, and surfactants in gel or other inert medium.

32 Solvent based paint strippers include aliphatic and aromatic hydrocarbons (such as petroleum distillates, mineral spirits, naphtha, benzene, toluene); chlorinated hydrocarbons (methyl chloride, ethylene dichloride, trichloroethane); N-methyl, 2-pyrrolidone (NMP); citrus terpenes; esters (dimethyl adipate, dimethyl glutarate, ethyl acetate); alcohols; and ketones.

33 Dwell times for coating removal products can be as long as 24 hours or more, depending on the coating removal product, substrate, and number and thickness of coating layers to be removed.

34 Quaternary ammonium chlorides are compounds found in disinfectants and anti-bacterial products for medical use as well as in building cleaning products.

35 Jokilehto, J., 'International Standards, Principles and Charters of Conservation', in Marks, S. (ed.), *Concerning Buildings, Studies in Honour of Sir Bernard Feilden*, Butterworth-Heinemann, Oxford (1996), pp. 55–81.

36 Normandin, K. C., Shotwell, L. B., and Stieve, D. R., 'Cleaning Atmospheric Pollutants and Contaminants from Masonry Surfaces: Modern and Traditional Methods', *Ninth North American Masonry Conference*, Clemson, South Carolina, 1–4 June 2003.

The White Tower and the Perception of Blackening

Carlota M. Grossi and Peter Brimblecombe

Abstract

A survey of the White Tower (at the Tower of London) by the authors assessed perceptions of blackening and the relationship between aesthetic damage and darkening of the stonework due to the deposition of soot. Two questionnaires were used to investigate perception, features, causes, and acceptability of blackening at two elevations (north and south-east) of the White Tower with different degrees of blackening. The first impressions of those surveyed were not so much of dirtiness, as of the grandeur or age of the monument. There was also evidence for a patina being seen as an indicator of antiquity or an element of building character among visitors. It is possible for the blackening of historic buildings to achieve a level of acceptability among visitors. Differences in colour appreciation at the two elevations suggested that it was possible to find a relationship between blackness of the stone walls and opinions that the building was dirty. The results of this work were compared with similar surveys of other monuments (e.g. cathedrals in Norwich, Milan, and Oviedo) and overall they hint that visitors hold fairly consistent attitudes towards dark coloration of building surfaces, which can help when making decisions about cleaning.

Introduction

The US Environmental Protection Agency (EPA) workshop 'Research Strategies to Study the Soiling of Building Materials' (1983) defined soiling as a 'surface degradation that can be undone by cleaning'.[1] However, in this paper we use the word blackening, rather than soiling, because we wish to make no assumptions as to whether the process is reversible or not. Blackening is represented by a darkening of the exposed surfaces through the accumulation of particulate matter and can be measured as a change in light reflectance, but here we are preoccupied with the way in which its perception lowers artistic worth, so raising issues of perception.

Research on the public perception of blackening of historic buildings needs to consider the 'aesthetics of soiling' and involves considering a wide range of social attitudes that influence public views about the appearance of buildings.[2] The perception of 'blackening' is thus complex and depends on the individual and on the general conditions of the local environment. Sometimes blackening of historical buildings represents a visual offence that becomes publicly unacceptable.[3] By contrast, on other occasions blackening can be aesthetically beneficial. There may be an expectation that old buildings should bear 'the grime of age'. Patinas are highly valued by many and have an aesthetic quality that often enhances the appeal of the building and that, particularly in the case of metals, can offer some protection. It has been argued that 'light or moderate blackening around architectural details could improve the visual appearance of the building by increasing contrast and enhancing shadowing effects',[4] while 'Heavy soiling eventually would lead to a uniform blackening, reducing the visual information or architectural details and completely obscuring the colour, texture and any shadowing effect.'[5]

The present investigation was a part of the European project 'Caramel' (Carbon Content and Origin of Damage Layers in European Monuments). Elements of this study focused on the 'aesthetics of blackening', where the main objective was to examine public perception of blackening of stone monuments and to establish a possible relationship between the air pollution and the appreciation of 'dirtiness'.[6] The research used questionnaires to examine visitor perceptions at different European sites. This paper summarizes the results of the surveys at a particularly significant building, the White Tower (at the Tower of London).

The White Tower is a key monument of great significance within European history. This ancient building occupies an urban site, but is now an open space free from immediate traffic, vehicles being an important contemporary source of blackening for urban buildings. Importantly it is called the White Tower, such that the use of the word 'white' raises public expectations and management concerns for those who have to maintain the façade in a way which meets these expectations. The way in which these expectations influence the public perception of blackening was a key part of our study of this building.

There is often a presumption that a 'blackened' building is one that requires cleaning, even though this has the potential to alter the historic character of the building. In the case of light-coloured stone buildings there is a very strong agreement between the visitor opinions that a building is dirty and the view that it needs cleaning.[7] However, when buildings are cleaned there is often a sense of loss. People who know a building well will often say after cleaning that it appears too light.[8]

Thin layers that accrete over time are widely known as patinas and these are seen to have both aesthetic and chemical characteristics. Their aesthetic value, which interests us here, derives from the fact that patinas do not alter the macroscopic nature of the object, but can enhance our appreciation of form. The way in which they do this is of course very subtle. Our study at the White Tower recognized the importance of patina and the loss that occurs during cleaning, and gave an opportunity to explore the balance between the desires on the part of the public both to have clean buildings and to maintain a patina of age.

Method

The site: the White Tower

The White Tower is situated in the central inner ward of the Tower of London. It was designed as a fortified palace and was constructed in the last decade of William's reign (1066–87). Although several types of stone can be found, the White Tower is mainly of Kentish ragstone. The French Caen limestone was also used.[9,10] The buttresses are of Portland limestone. The Kentish ragstone is a sandy glauconite limestone from the Lower Greensands of early Cretaceous age. It is dark blue to greeny-grey in colour. The amount of silica in some beds rises to the level where these beds are regarded as sandstones and not limestones. Caen stone is a Middle Jurassic fine-grained limestone, yellowish to yellow-white in colour.[11] Portland stone is an oolitic, creamy-whitish Upper Jurassic limestone with different amounts of shell.

The questionnaires

Two questionnaires examining the perception of the blackening of buildings were carried out on site at different historical European buildings by the authors and collaborators (see acknowledgements). These monuments presented a range of blackness and occupied different surroundings in terms of open spaces, nearby roads, historical importance, environmental quality, etc. The sites include several cathedrals such as Norwich (UK), Oviedo and the basilica of Llanes (Spain), and Florence and Milan (Italy), along with a few other significant buildings of a more mercantile character.[12]

The questionnaires were intended to be short and easy to answer, such that passers-by were willing to be engaged. They consisted of questions about impression of the building, surroundings, sensation of dirtiness, apparent causes of blackening, and feeling of colour. The first questionnaire was undertaken in the late summer of 2002. The second questionnaire – planned on the basis of the results of the first questionnaire – was undertaken during the late summer and early autumn of 2003.

The first questionnaire (Qa) asked the following questions:

1 What words best describe your impression of the appearance of this building?
2 Do you think this building appears dirty? What is it about the appearance of this building that makes it seem dirty to you?
3 Is the surrounding environment appropriate for this building and why?
4 In general, when you look at any building what is it about the building that makes it appear dirty?
5 From your perspective what are the main reasons for buildings becoming dirty?
6 What colour would you say best describes the building?
7 Which colour on this chart characterizes best the appearance of the building?

The chart referred to in Question 7 was the Munsell[TM] Colour Chart,[13] which includes the dimensions of 'hue', 'value', and 'chroma' and assigns letters and numbers to colours. Hue denotes the colour (red, yellow, green, etc.) and forms the colour wheel. Value is the lightness of the colours (light, dark), and chroma refers to colour saturation (vivid, dull). As might be expected, value showed the strongest relation to public perception of dirtiness. There was less clarity in the choice of chroma and hue.[14]

The second questionnaire (Qb) contained similar questions, but they were designed to focus more on issues that had emerged in Qa. The questions were as follows:

1 What words best describe your impression of the appearance of this building?
2 What does this building represent for you?
3 What word best describes the surroundings?
4 Do you think this building appears dirty? Why?
5 Is this building light or dark?
6 What source of pollution do you think could make this building dark?
7 Do you think the building needs cleaning?
8 Which colour on this scale best characterizes the appearance of the building?
9 Which colour on this scale should this building be?

The scale mentioned in Question 9 is a grey scale of twelve elements – from white to almost black. Both the grey scale and the Munsell[TM] Colour Chart values were converted to standard values of reflectance or lightness

Figure 1 Two views of the White Tower, where the questionnaires were taken: SE corner (left) and N elevation (right).

L* (system L*a*b*) by measuring the charts with a Minolta colorimeter CR-300, thus reducing all data to a reasonably familiar parameter.[15] An L* of 100 % is pure white. These L* values were taken as representative of the perceived lightness of the building (not a measurement of the reflectance of the stone in the building) and will be designated L_p in this paper.

The first questionnaire was run in September 2002 at two locations at the White Tower, the first looking at the south-east corner (SE corner) and the second looking at the north elevation (N elevation) (Figure 1). From the first site it is possible to see simultaneously both the south (S) (lighter/restored) and the east (E) (darker) elevations. The second site (N elevation) reveals a darker façade that is placed opposite to the Waterloo Barracks (which appear quite clean). The later questionnaire was carried out at only the N elevation in October 2003.

The sample

The sample consisted of 100 people at both the SE corner (Qa) and the N elevation (Qb) and eighty-four people at the N elevation (Qa). The questionnaires were addressed to individuals, couples, and families from countries all around the world, although mainly UK and Europe, USA, and Australia–New Zealand.

Results and discussions

The answers are summarized in Table 1. The first impression created by the appearance of the White Tower (questions 1a/b) is 'magnificent, beautiful, nice, I like it, etc.' (40–50 %), followed by 'old, ancient, or historic' (20–35 %) (Qa and Qb). The unprovoked or unprompted comments that the façade was dirty were rather infrequent (less than 5 %), but occasionally there were oblique remarks, such as 'It's not white'. Without prompting, blackening is not a dominant first impression of the appearance of ancient historic buildings.[16]

When prompted (questions 2a, 4b) the percentage of respondents arguing 'It's dirty' rises to 20 % at the lighter SE corner and 39 % (Qa) and 34 % (Qb) at the darker N elevation. The difference between the SE and N sites seems to be significant. The darkness of a façade clearly influences beliefs that it is dirty.[17] Thus predominant reasons for finding the building dirty (questions 4a/b) are those related to colour or darkness, and as noted the expectation about its colour which led to the view 'It's not white'. Contrast in colour was a frequent answer at the SE corner, where people could see both the cleaner S elevation and the dark E elevation. The answer 'soot, pollution' is also common at the N elevation. It is especially notable that the reasons for not finding the building dirty are mainly those of antiquity: 'It's old, ancient, naturally aged …'. Somebody even suggested it had been affected by the 'patina of time' and sometimes the answer is 'It's

	SE corner (Qa)	N elevation (Qa)	N elevation (Qb)
First impression	1. Magnificent, impressive, nice, beautiful… (40–50%) 2. Old, ancient, historic, medieval… (20–35%)		
Surroundings	Appropriate (~ 90%)		Positive evaluation (> 95%)
Is it dirty?	No (80%)	No (61%)	No (66%)
Needs cleaning?	–	No (62%)	
General features / Why is/isn't it dirty	Colour (47%)	Colour (73%)	Not dirty: Old Dirty: Colour
Causes / type of pollution	Pollution (63%)	Pollution (75%)	Traffic (~ 40%)
Visual colour / light/dark	Cream	Grey	Diverse
Munsell™ colour	White*	Grey	–

* Mainly S elevation (probably few of the E elevation)

Table 1 Main answers for questionnaires with details of percentage of visitors making given responses.

not dirty, it's old'. In general, the percentage of visitors finding the building was 'not dirty' seems to be in accordance with the percentage of visitors finding the façade 'light coloured' (question 5b) (Figure 2). The most common word assigned to the colour of the building (question 6a) was 'cream' for the S elevation and 'grey' for the N elevation. The mode for the selected Munsell™ hue, value, and chroma was '2.5Y 8/1 White' in the S elevation, while it is more variable in the N elevation, but resulting in the colour 'Grey' (question 7a).

Finally, the number of visitors answering that the building 'does not need cleaning' (question 7b) at the N elevation was 60%, and seems to be significantly higher in the group of people answering 'not dirty'. In both cases (Figure 3) the chi-square p is <0.001 when cross tabulating the questions with perception of dirtiness.

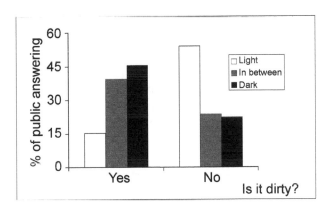

Figure 2 (left) Percentage of public seeing the façade as light dependent on their answer it is 'dirty' or it is 'not dirty'(N elevation, Qb).

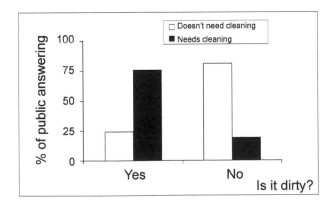

Figure 3 (right) Percentage of public seeing a need for cleaning dependent on their answer it is 'dirty' or it is 'not dirty'(N elevation, Qb).

Dirtiness and perceived reflectance

The averaged perceived lightness is 72 % and 59 % respectively for the S and N elevations in the first questionnaire (Qa) (Table 2). These differences in the appreciation of colour of the two elevations suggest that there is a relationship between the perceived blackness of the stone walls and the perception of 'dirtiness'.

The results from the second questionnaire (Qb) are in agreement with this idea. The perceived lightness (L_p ~ 61 %) is similar to those given at the N elevation in Qa. However, the average values given for desired lightness (L_d) seem to be comparable to those of the S elevation (Table 2). Moreover, some Qb respondents chose the lightest grey scale value offered ('equivalent' to L_p near 90 %) because they said 'It is the White Tower'.

Figure 4 shows that where people believe a building is dirty they will desire it to be very much lighter than when they perceive the building to be 'not dirty'. This has some management importance, because as long as people perceive a building 'not dirty' (i.e. 'it is old or historical'), they will tend to be relatively happy with its lightness or tone (Table 3).

There also appears to be a relationship between the value of perceived lightness L_p and the appreciation of dirtiness as shown in Figure 5. However, if we compare data from the N elevation of the White Tower with results obtained at other historical sites using the same questionnaire, we note that despite a relatively dark colour fewer people regard it as dirty than at similar sites (see Figure 5). It is as if the special significance, age, or character of the White Tower allows visitors to accept its dark patina.

These results agree with some of the outcomes of a study reported by Andrew,[18] who found that parameters such as pleasing colours, cheerful, tidy, well looked after, are related to relative cleanness. In addition Andrew found a general tendency to estimate the age of soiled buildings to be older than cleaned counterparts. This seems to agree with the sense of the public response 'It's not dirty; it's old' in the Tower questionnaires.

Site	Questionnaire	Sample size	Reflectance (L %)	
			L_p or L_d	Mean (stdev)
SE corner	a	95	perceived	72/67* (±12/13)
N elevation	a	83	perceived	59 (±15)
N elevation	b	96	perceived	61 (±9)
N elevation	b	94	desired	73 (±12)

*It varies depending on which aspect people view, with the higher value mainly referring to the S elevation and the lower value referring to the SE corner as a whole.

Table 2 Perceived and desired reflectance at the surveyed sites at the White Tower.

Desired reflectance or lightness	Total public choices	Public answering 'Dirty'	Public answering 'Not dirty'
Lighter	54	27 (90%)	27 (44%)
Darker	4	1	3
Ties	34	2	31 (51%)

Table 3 Difference between perceived and desired lightness or reflectance at the N elevation in questionnaire b (question 9) with details of percentages in brackets.

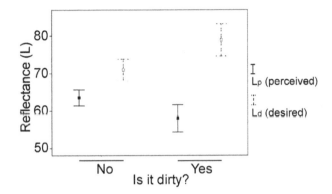

Figure 4 Perceived (L_p) and desired (L_d) lightness or reflectance given by people answering 'dirty' and 'not dirty' at the N elevation (Qb).

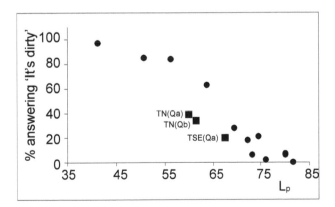

Figure 5 Percentage of public answering 'it is dirty' versus the Munsell™ value (lightness); questionnaires a and b used at different European historical buildings. (Square markers represent White Tower sites.)

While it would be interesting to be able to relate perceived colours of a building to measurements made with a colorimeter, this is very difficult.[19] Most particularly the colorimeter takes measurements from a small area of the surface, while human perception seems to integrate the colour from a large part of the façade.

Concluding remarks

Analysis of the responses shows that the public's first impression of the White Tower was not of dirtiness, but of the grandeur or the age of the monument. There was also evidence for a sense that the dark patina may be an emblem of antiquity or character. It is possible for the blackened surfaces of historic buildings to achieve a level of social acceptability.

Differences in colour appreciation at the two elevations, and from comparisons with similar surveys of other monuments, showed it was possible to find a relationship between blackness of the stone walls and perceptions of 'dirtiness'.

The name 'White Tower' can influence some public perceptions of the building. The survey reveals that the name, the context, the architecture, and the age are interpreted by the public in ways that affect their views of blackening. People seem more able to accept the White Tower to be blackened than other light-coloured stone buildings of historic value, despite the fact that it is called the White Tower. Thus management decisions about cleaning have to confront a complexity of visitor feeling and engage public acceptance of the notion of historic patina.

Decisions to clean are usually based both on a concern about physical damage and on an assumed public preference for clean buildings. However, it is rare for public opinion to be sampled prior to cleaning, so there is little to guide managers in assessing the point at which this visual nuisance becomes unacceptable. Public perceptions are not easy to integrate with considerations such as available finance, concerns over physical damage, or the importance of having a clean building for specific events or celebrations. Nevertheless, better understanding of the public perception of blackening would allow managers to make consistent and perhaps more justifiable decisions about cleaning.

Biography

Dr Carlota M. Grossi
Carlota is a senior research associate at the University of East Anglia (UEA). Her speciality is building stone decay and conservation.

Professor Peter Brimblecombe
Peter is a professor of atmospheric chemistry at UEA with a particular interest in the relationship between air pollution and cultural heritage.

Acknowledgements

This paper has benefited from work within the CARAMEL project (ENV4-CT-2000-0002) funded by DG Research of the European Commission. We also wish to acknowledge Ms Jo Thwaites and Mr Richard Roberts of Historic Royal Palaces, who assisted our work at the Tower of London; to Professor Cristina Sabbioni, Dr Alessandra Bonazza; Ms Jessica Zamagni, and Mr Alessandro Sardella, of the Instituto di Scienze dell'Atmosfera è del Clima (Bologna, Italy); and Mr Matthew Hassall, Ms Begoña Sastre, and Dr Derek Bowden, of the University of East Anglia for assisting with the questionnaires. We are also grateful to Ms Louise Bacon of the Horniman Museum for the loan of the Minolta Colorimeter.

Notes

1   Haynie, F. H., 'Theoretical model of soiling of surfaces by airborne particles', in Lee, S. D. et al. (eds.), *Aerosols*, Lewis Publishers, Chelsea, Michigan (1986), pp. 951–9

2   Ball, J., Laing, R., and Young, M., 'Stone Cleaning: Comparing Perceptions with Physical and Financial Implications', *Journal of Architectural Conservation*, Vol. 6 No. 2 (July 2000), pp. 47–62.

3   Hamilton, R. S. and Mansfield, T. A., 'Airborne Particulate Elemental Carbon: Its Sources, Transport and Contribution to Dark Smoke and Soiling', *Atmospheric Environment*, 26 A (18) (1992), pp. 3291–6.

4   Andrew, C., 'Towards an Aesthetic Theory of Building Soiling', in *Stone Cleaning and the Nature, Soiling and Decay Mechanisms of Stone*, Donhead Publishing, Shaftesbury (1992), pp. 63–81.

5   Grossi, C. M. and Brimblecombe, P., 'Aesthetics of Simulated Soiling Patterns on Architecture', *Environmental Science and Technology* 30 (2004), pp. 3971–6.

6   Brimblecombe, P. and Grossi, C. M., 'Aesthetic Thresholds and Blackening of Stone Buildings' in *The Science of the Total Environment*, in press, available online 13 March 2005 www.sciencedirect.com.

7   Ibid.

8   Grossi, C. M. and Brimblecombe, P., 'Aesthetics and Perception of Soiling', in *Air Pollution and Cultural Heritage*, Balkema (2004).

9   Leary, E., *The Building Limestones of the British Isles*, Building Research Establishment Report, Department of the Environment (1983).

10 Dimes, F. G., 'Sedimentary Rocks', in Ashurst, J and Dimes, F. G. (eds.) *Conservation of Building and Decorative Stone*, Butterworth Heinemann, London (1990).

11 Ibid.
12 Grossi, C. M. and Brimblecombe, P., 'Aesthetics and Perception of Soiling', in *Air Pollution and Cultural Heritage*, Balkema (2004).
13 *Munsell*^{TM} *Colour Charts* (revised edition) are currently available from GretagMacbeth, New York, a subsidiary of Amazys Holding AG, Regensdorf, Switzerland (1994).
14 Brimblecombe and Grossi, *op. cit.* (note 6).
15 Ibid.
16 Ibid.
17 Ibid.
18 Andrew, *op. cit.* (1992).
19 Brimblecombe and Grossi, *op. cit.* (note 6).

Sydney Opera House

Analysis and Cleaning of the Concrete

Paul Akhurst, Susan Macdonald and Trevor Waters

Abstract

This paper discusses the cleaning and conservation of the folded concrete beams, a key architectural feature of Sydney Opera House. This cleaning and conservation work attempts to develop an appropriate method and specific techniques for dealing with the exposed concrete that can be applied to other areas of the building in the future. It describes how the development of the methodology relies on a sound understanding of the historical development and construction of the building and the significance of the concrete in the architectural language of the place. It exemplifies the relationship between the Conservation Plan *and the* Utzon Design Principles *that together provide the framework for decision-making at Sydney Opera House and shows these documents working in practice.*

Through interviews with labourers, supervisors and engineers, consideration is given to site practices employed in 1964 that have proved to be stable and long-lasting. The development of the approach to the cleaning of the concrete also shows the value of combining oral and published history with laboratory assessment of conservation techniques.

Introduction

This is the second paper relating to the conservation of Sydney Opera House.[1] Previously the framework for the ongoing conservation and care of Sydney Opera House was described. This includes the implementation of a *Conservation Plan*,[2] together with the *Utzon Design Principles*,[3] which jointly provide the guiding policy for the ongoing management of this important building. These policy documents have been adopted by the Sydney Opera House Trust as its managers, and are also used by the

Figure 1 Sydney Opera House was envisaged as a sculptural element in a magnificent harbour setting. It consists of a substantial, solid platform that draws on historic references. Atop the platform are the shells, constructed as segmented precast concrete ribs, overlaid with glittering white glazed tiles. (Courtesy Max Dupain & Associates)

government heritage agencies responsible for overseeing the conservation of the building.

The use of extremely high-quality exposed concrete at Sydney Opera House, cast *in situ* and incorporating precast elements, exploited the organic nature of the material – its colour intended to refer to nature. As Utzon said, 'You find a similar situation in Gothic cathedrals, where the structure is also the architecture.'[4] The top surface of the shells, in contrast, are covered with white glazed tiles, with the exposed structural precast concrete ribs forming the ceiling below (Figure 1).

The exposed concrete folding beams are a key feature of Sydney Opera House, providing the means by which the huge spans could be provided without columns, but also contributing to the architectural character and quality of the place. The building's architect Jørn Utzon described the beams as 'these ribs shaped so they elegantly express the forces within the structure. They express the harmony within the structure.'[5]

The folded beams support the southern part of the platform on which sit the soaring shells of Sydney Opera House. The beams rise from the ground, support the steps, and provide a concourse beneath. The folded beams also form the ceilings of the interior box office spaces, the preliminary foyers to the Opera Theatre and the Concert Hall, and the ceilings of the grand external anteroom to these foyers – the Vehicular Concourse.

The *Conservation Plan* and the *Utzon Design Principles* both include detailed policies for the exposed concrete generally and the folded beams specifically, which are identified as being of 'exceptional significance'.[6] The work discussed in this paper includes cleaning and conservation of the folded exposed concrete beams that encase the Utzon Room and Vehicular Concourse below the monumental steps of the podium.

The recent concrete conservation and cleaning works to the Utzon Room and Vehicular Concourse required careful analysis and consideration. The importance of the aesthetics of the concrete finish in these areas demands that any conservation work recognizes this and that it reverses the effect of time and pollution on the finish of the concrete to preserve the material authenticity of the building and retard the environmental effects that contribute to concrete decay.

The study described in this paper attempts to address Sydney Opera House architect Jørn Utzon's desire to 'conceal the defects and bring it up to a uniform and acceptable standard',[7] and the *Conservation Plan*'s Policy 41.1 that 'beams should remain unpainted and their detail unobscured'.[8] Contemporary accounts and photographs highlight that the soffits of the folded concrete beams were blemished by a white residue when the formwork was removed in 1962, well before leaks through the protective membrane that sits above the folded beams became an issue. Thus the goal of determining what is 'clean' encompassed the removal of polluting agents and recapturing the visual continuity absent from much of the original surface. The initial investigation work described in this paper provides an overall philosophy and framework for the whole process of conservation from the initial consideration of the need for cleaning and conservation through to the completion of repairs and ongoing maintenance of the structure.

The folded concrete beams

Background

Jørn Utzon's original competition drawings of 1956 show beams spanning the width of the lower concourse, which forms access for vehicular traffic and pedestrians into the podium (Figure 2).[9] The folded beam profiles were arranged in a smooth flowing pattern, starting off as a 'T' section and gradually altering into an open 'U' profile where the tensile stress is concentrated in either the soffit or the crown. The geometry of the folded beams is a conic section in elevation,[10] entirely created with straight lines in profile (Figure 3).

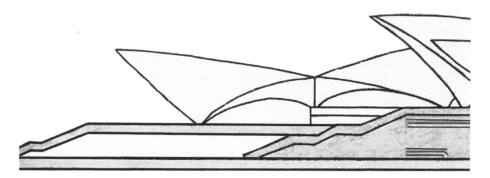

Figure 2 The supporting concrete beams span 47.5 metres (156 feet) across and down the slope of the grand stairs, without columns. The top surfaces of the structural beams are covered by precast granite paving, which incorporates open joints to let surface water run off into troughs formed between the beams. 'Submission no. 218', SOH Competition Drawings (Dec. 1956). Technical Information Management Services, Record No. D1419. (Courtesy of Sydney Opera House Trust)

Figure 3 Detail of the folded beams. (Trevor Waters, 2004)

The construction of Sydney Opera House commenced on site on 5 May 1959. Late in 1960 the first pair of folded concrete beams was poured. Ove Arup and Partners were responsible for the design and construction supervision of the folded concrete beams. Despite the rigorous contractual requirements for the formwork, the beams did not achieve the smooth lines and sharp arises necessary to create the strong geometry that Utzon sought. The narrow formwork (Figures 4 and 5) precluded the efficient vibration of the concrete, particularly at the U-shaped knee. Excessive slurry escape had indicated formwork movement, whilst an obvious sagging at the knee indicated a fault in the supporting scaffold.[11] Despite efforts by the site supervisory staff to control the quality of the formwork,[12] the soffit of the folded slab acquired streak marks from the slurry escaping (Figures 6 and 7).[13]

Figure 4 Constructing the formwork for the folded beams. The sanding down of the grounds was insisted upon, thus weakening the 4 inch x 1 inch boards and so restricting the number of reuses, 1962. (Courtesy Max Dupain & Associates)

Figure 5 Large sheets of 3/16 inch plastic-faced plywood were used to line the formwork for the folded beams. Even minor damage involved the loss of the whole of that sheet, 1962. (Courtesy Max Dupain & Associates)

Figure 6 The special formwork called for by Ove Arup and Partners involved sinuous curves and elaborate setting out procedures for the soffits and beam sides. The tolerances demanded by the engineer were more consistent with precision joinery of cabinet making rather than formwork, 1962. (Courtesy Max Dupain & Associates)

Figure 7 This 1962 photograph shows the beams shortly after the formwork was stripped and reveals that the cleaning out of the forms was unsatisfactory. Excessive slurry escape in beams K1/K2 indicated formwork movement. The surface finish of the concrete was also unsatisfactory. (Courtesy Max Dupain & Associates)

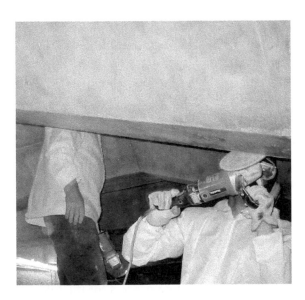

Figure 8 The surface of the beams was burnished to create a uniformly white surface. (Trevor Waters, 2003)

The contractor developed a method of treating the freshly stripped concrete, in the northern foyers, to provide a more even finish. The concrete beams were rapidly dried with fan-forced heaters and slowly buffed with a sanding disc to disperse the surface efflorescence. The concrete was then burnished with coarse pads, on variable speed drills, to make the surface uniformly white (Figure 8).[14] The quality of finish met with Utzon's approval, but he left Australia before treatment of a similar standard could be found for the hardened folded concrete beams.

The 2002 *Utzon Design Principles*, which specifically mentions the concrete finish, confirms that a cleaning and resurfacing process is still necessary which is reversible and does not obscure the texture of the concrete:[15,16]

> Some form of treatment of these surfaces as a whole, or in part will be necessary to conceal the defects and bring it up to a uniform and acceptable standard. Such a standard has now been set by the quality produced in the shell structure and most recently in the bar areas, but the actual treatment will depend on the outcome of experimental work ...

Defining an approach to cleaning of the exposed concrete finishes

The condition of the folded beams has deteriorated through prolonged penetration due to membrane failure[17] and the common practice of direct fixing to the beam structure, as well as *ad hoc* service penetrations. As required by the *Conservation Plan* and identified in its policies,[18] careful investigation and analysis of the surface was undertaken. The intention of this approach was to avoid further deterioration and to develop potential conservation treatments that would be stable and durable over time. Cleaning treatments should remove contaminating matter as well as even out the mottled concrete substrate that Utzon regarded as defective and detrimental to the intended character of the original surface.

The *Conservation Plan* contains a number of detailed polices about the exposed concrete beams. These include policy on surface finish (colour and texture), the importance of using tried and tested methods and proper diagnosis of problems, technical adequacy of repairs, and the retention of construction markings or characteristics. Likewise, the *Utzon Design Principles* also addresses the concrete finish in terms of colour and texture.[19]

The goal of the cleaning process is to rectify obvious defects and also to establish the appropriate methodology and techniques for the cleaning and conservation of other sensitive areas of off-form concrete ceilings within Sydney Opera House.[20] The objective of the cleaning process for the folded concrete beams is to:

1 remove or reduce pollutants from the fabric whilst retaining the natural visual properties of the concrete surface;
2 treat the unpainted concrete soffits to even out blemishes and to recapture their early visual appearance in texture, colour, tone, and uniformity;[21]
3 ensure that the cleaning procedure above is only executed through processes that are fully tested on the actual fabric and for which the consequences are known, and not to introduce agents that may result in future visual change or deterioration;
4 retain the original post-tension set-out lines, vibrator markings, and patches that help to explain the past history of the fabric; and
5 involve processes that are stable, long-lasting, and as simple in application as the complexity of the task permits.[22]

Field survey and testing

Prior to commencement, a photographic record of the areas to be cleaned was developed, with the images compiled to provide detailed mapping of the cracks and water ingress. Stress cracks radiating from the anchorages that occurred during construction were identified from the resident engineer's records (Figure 9).[23]

Some water staining had occurred along the cracks, which suggested that the membrane covering the slab had failed. There was evidence of water ingress. Leaching was noted at a crack in the knee and a scraping was taken. Preliminary analysis indicated that the material is likely to be calcium carbonate with low levels of sulphate. Further scrapings were taken for laboratory analysis to determine the nature and source of the blemishes.

Three concrete core samples were analysed to determine chloride and sulphate penetration profiles at different depths below the surface of the concrete. At selected depth intervals, layers were ground in a laboratory disc mill to less than 1 mm and analysis was performed.

Table 1 and Table 2 illustrate that with the exception of some results from the outer layers, the chlorides were well below the 1150 ppm (0.4 per cent by weight of cement) threshold of concern,[24] and would not constitute

Figure 9 A photographic survey of the folded slab to the Utzon Room mapped the open crack and areas of water ingress. The arrows highlight the crack and leaching of the beam as a result of water penetration. (Trevor Waters, 2004)

Depth (mm)	Chloride (ppm)	Sulphate (ppm)
All	100	110
0–10	100	1,020
20–30	70	770
40–50	50	790

Table 1 Chloride and sulphate concentrations at depths up to 50 mm within the Utzon Room.

Depth (mm)	Chloride (ppm)	Sulphate (ppm)
All	580	2,050
0–10	770	420
20–30	130	340
40–50	<10	450
140–150	10	330
160–170	<10	360
180–190	20	910

Table 2 Chloride and sulphate concentrations at 190 mm depth within the Vehicular Concourse.

a significant risk of chloride-induced corrosion of the reinforcing steel with cover greater than 20 mm. It also appears that the original mix of concrete had a very low chloride content, in the order of 20 ppm. While the absolute levels of sulphates detected remained well below the threshold (14,000 ppm or 5 per cent by weight of cement), these results indicate some elevation of sulphates above the background.[25]

Prolonged exposure to the sulphates and chlorides in sea spray may have promoted the development of insoluble calcium sulphate dihydrate (gypsum) and calcium chloride to the top and lower surface of the beams in the Vehicular Concourse areas. There is no evidence of salt penetration through the concrete, or of internal seawater attack.

Surface scrapings of the folded concrete beams were collected from different locations within the Utzon Room to determine the composition of the stains and blemishes to the soffits.[26] The laboratory results are graphically represented as a percentage by weight of calcium carbonate $CaCO_3$, calcium hydroxide $Ca(OH)_2$, hydration water H_2O, organics, and quartz (or other) (Tables 3 and 4).[27]

Thermal analysis of the concrete surface scraping samples reveals a greater proportion of mineral deposits than of organic matter such as

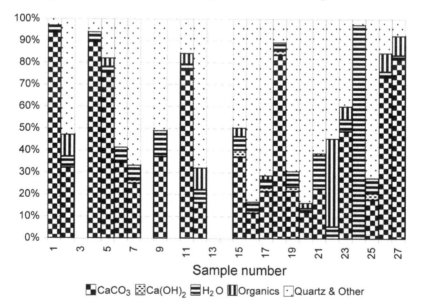

Table 3 Analysis of surface scrapings.

CaCO$_3$	Ca(OH)$_2$	H$_2$O	Organics	Quartz and other
41.50%	0.67%	9.58%	4.64%	43.61%

Table 4 Scrapings average % composition.

algae, fungi, bacteria, smoke residue, or formwork oil. This type of organic deposit is more likely on the exposed beams at the extremities of the Vehicular Concourse. Here the weathered and porous surface presents an ideal habitat for organic growth to establish.

Notable quantities of calcium hydroxide $Ca(OH)_2$ were detected in samples collected from active water leaks. Samples 15 to 19 were taken from areas with distinct white efflorescence close to construction joints or cracks with a stream of deposits on the soffit. The weeping of lime is usually caused by acidic rainwater (carbonic acid) dissolving calcium from the cement matrix to form calcium bicarbonate $Ca(HCO_3)_2$. Precipitating calcium salts form as a bloom or indistinct whitening around the leak.[28]

The ratio of $CaCO_3$ to quartz over the soffit of the concrete is variable. However, there is no evidence of significant leaching of concrete components and subsequent salt deposition on the soffit of the beams. The variable coloration is due to the uneven distribution of calcium hydroxide $Ca(OH)_2$. This is a natural by-product of the hydration of cement that has carbonated on exposure to air forming insoluble calcium carbonate $CaCO_3$. This chalky residue is the most likely cause of the 'cloudiness' in the matrix of the concrete and streaks, clearly described by the resident architect and evident in the contemporary photographs taken during construction (See Figure 7).

Further chemical analysis of surface scrapings was undertaken to determine if there was a compositional variation between the staining in the exposed Vehicular Concourse and the interior of the Utzon Room (Tables 5 and 6).[29] The expected outcome was that due to the failure of the original membrane there would be evidence of external seawater attack and salt penetration in both areas sampled.

Sample 1A is taken from the level portion of the beam adjacent to the south-eastern wall of the Utzon Room. Serious leaking of water through the open crack has lead to the formation of stalactites on the soffit of the folded beam. There is evident failure of the original and new membranes.

The scraping was taken from this fissure and analysis has proved it to contain predominantly calcium salts (5.3 per cent). This is symptomatic of a significant rainwater leak. The results show the composition of magnesium oxides to be very low, less than 0.5 per cent, with a small percentage of both sodium and potassium oxides. These low concentrations do not indicate a more severe problem of saltwater migration that is having an adverse effect on the concrete durability. Few sulphates were detected (0.03 per cent) and the cement surrounding the crack does not have the friable material that would be expected if the crack was caused by the softening or dissolution of the hydrated cement by sulphate present in seawater. Furthermore the low levels of iron oxide suggest that rusting of the reinforcement by water ingress may not be widespread.

	Sample 1A Utzon Room	Sample 5A Vehicular concourse
	Raw element analysis results (% by weight)	
%Fe	0.2	2.9
%Ca	3.8	2.6
%Mg	0.1	1.2
%Na	0.8	0.9
%K	0.8	1.0
%SO$_3$	0.03	5.7
%Cr	n.d	n.d
	Recalculated oxide composition (% by weight)	
%Fe$_2$O$_3$	0.3	4.1
%CaO	5.3	3.6
%MgO	0.2	2.0
%Na$_2$O	1.1	1.2
%K$_2$O	1.0	1.2
%SO$_3$	0.03	5.7
%Cr	n.d	n.d
Oxide Σ	7.8	17.9

Table 5 Chemical analysis of surface scrapings.

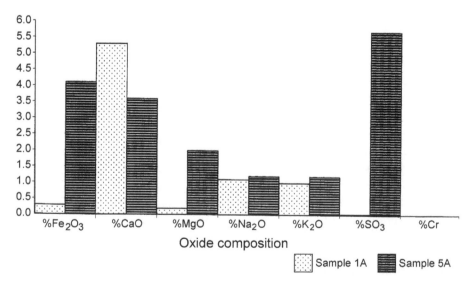

Table 6 Graph of the recalculated oxide composition (% by weight) for the Utzon Room (Sample 1A) and the Vehicular Concourse (Sample 5A).

Figure 10 The folded concrete beams to the Utzon Room retain the original mottled and rust-stained finish that was evident in 1962 construction photographs. (Trevor Waters, 2003)

Sample 5A was collected from the sloping soffit of the western beam in the Vehicular Concourse. Although the internal concentration of sulphates was low, the superficial build-up of SO_3 (5.7 per cent by weight) is a concern. Unless sulphates being deposited by sea spray are regularly cleaned off, there is a risk that they will attack the cement matrix and induce corrosion of the reinforcing steel. Very slight sulphate attack has been detected on the external faces of the first and last beams, which are those closest to the harbour.

The 4 per cent iron oxide component is consistent with an overall rust coloration within the Vehicular Concourse and is probably a stain from construction. The formwork was exposed for long periods, largely due to the extra quantity and complexity of reinforcement, causing excessive weathering deterioration. The increased reinforcement made it difficult to clean out the forms before casting.[30] Rust from the bars was washed onto the soffit formwork by rain, resulting in the permanent staining of the concrete face (Figure 10).

The preceding test results support the view that the concrete is sound, of high strength and relatively impervious to moisture. Furthermore the permeability of this high grade of concrete would not allow a ready migration of salt except through obvious cracks.[31] The analysis of cores and surface scrapings has demonstrated that the current efflorescence and staining was present, in part, when the slabs were first poured. The discontinuous and uneven surface that so displeased Utzon in 1962 is integral to the structure and is not the result of salt penetrating through the folded concrete beams and depositing on the surface.

Conservation programme

As recently as 2002 Utzon continued to express his dissatisfaction with the presentation and finish of the folded concrete beams to the Vehicular Concourse, Utzon Room, and southern foyers:[32]

One way to remedy this is to raise the light level in the area artificially. Another way could be to whitewash the concrete surfaces of the ceiling above the area. Whitewash can be cleaned off again, or applied in such a way that it does not camouflage the concrete texture. Trials in 'selected' areas would be needed to establish the correct procedure.

Steve Tsoukalas, conservation tradesperson, has worked on the construction and maintenance of Sydney Opera House for thirty-five years. He retained the tools necessary for the original finishing process and was able to demonstrate the original buffing process that turned dull, blotched concrete into a lustrous, translucent finish (Figure 11). On freshly stripped concrete the cleaning process is similar to the process employed in the buffing of the northern foyer beams after 1964. The surface staining, rust streaks, and calcium hydroxide $Ca(OH)_2$ blemishes are dispersed evenly with a coarse pad on a 7-inch variable-speed buffing tool. The precipitate is worked to a constant edge in a tracking motion rather than broad uneven strokes, taking care not to leave behind swirl marks or a slurry build. The pressure is gentle and the speed is slow (400 rpm). As the calcium hydroxide is buffed the slurry becomes a tacky or viscous film on the surface. Within a month the surface is fully carbonated to a translucent

Figure 11 Steve Tsoukalas demonstrates the use of the original drill used to burnish the soffits of the freshly stripped concrete. The hardened concrete is currently 'reactivated' with a calcium bicarbonate solution that evens out the blemishes and fills the pores, creating a denser surface. This figure demonstrates the process only – correct safety procedures were followed when undertaking the task. (Trevor Waters, 2003)

calcium carbonate film that has a strong polarizing effect when viewed through a polarizing film. This is typical of calcite crystal formation.[33] The same tools and action were initially unsuccessful when used for the conservation treatment of hardened concrete. Site trials demonstrated that once the calcium hydroxide (lime) had fully carbonated on the folded concrete beams, the resultant calcium carbonate (calcite) was difficult to remove. As calcium carbonate is virtually insoluble and has little binding power it is important that it is dissolved and dispersed evenly into the micropores of the concrete. An emulsion was required that would significantly increase the solubility of calcium carbonate in water. Laboratory experiments and site trials investigated lime washes and thick calcium carbonate suspensions that were 'painted' over the surface, but obscured the character of the concrete. Sugar solutions readily formed soluble calcium saccharate; however, analysis confirmed that even after repeated washing residual sugar formed loose organic bonds with hydroxides, reducing alkalinity and increasing the risk of degradation. Sodium silicate, which reacts with free hydroxides to form a glass-like binder around the calcium carbonate, was also considered.[34] These treatments were rejected as they did not fulfil the primary objective of involving processes that are stable, long-lasting, and simple in application.

Conservation treatment

The treatment adopted follows the same process that enables acidic rainwater to dissolve calcium carbonate from within cracks and disperse it evenly over the soffits of the beams. Thus the hardest problem, water ingress, becomes the basis of the simplest solution. The mixture is easily prepared by dissipating calcium carbonate in a bottle of carbonated water. This carbonic acid H_2CO_3 dissolves the insoluble $CaCO_3$ into the soluble calcium bicarbonate $Ca(HCO_3)_2$.

$$Ca^{2+}(aq) + (HCO_3^-)_2(aq) <> CaCO_3(solid) + CO_2(gas) + H_2O(liquid)$$
calcium bicarbonate < > calcium carbonate + carbon dioxide + water

The soluble 'life' of the bicarbonate is limited. If the container is left open CO_2 gas is desorbed and $CaCO_3$ precipitates from solution; the same effect is obtained at drying. This process when applied to cleaning hardened concrete has the potential to induce controlled dissipation.

The substrate is first steam-cleaned to remove organic deposits and soluble salts, leaving the micropores more open. The substrate is left to dry for several hours so that it will absorb the carbonated water faster and deeper. A fresh bicarbonate solution (prepared just before application) is applied as a thin wash over the surface of the concrete. The depth to which the calcium bicarbonate will penetrate depends on factors such as opened porosity or moisture content.

Once the calcium bicarbonate is applied and absorbed into the pores of the concrete, the subsequent evaporation of the water will generate precipitation of the calcium carbonate into the pores. Slow, natural evaporation is preferable to accelerated drying. Gentle buffing with a coarse finishing pad, during drying, changes the surface from a translucent film to a tacky opaque glaze. This is the same effect as observed on freshly stripped concrete. When it is dry a microscopic layer of calcium carbonate is formed on the surface and within the pores of the concrete. This treatment evens out the dappled blotches, disperses the rust stains, lightens the surface, and recaptures the character of the concrete beams as they appeared after removal of the formwork (Figure 12).

Consequently, Sydney Opera House has adopted a low-pressure cleaning solution saturated with calcium and containing an excess of dissolved CO_2. Carbon dioxide is bubbled through a suspension of water and calcium carbonate to form an acidic calcium bicarbonate solution (6.75 pH at 11°C and 140 psi). Calcium solids are initially dissolved from the surface of the concrete and then evenly deposited as the water evaporates and carbon dioxide is lost from the liquid. In practice the bicarbonate solution removes superficial grime and efflorescence, evening out the white staining (Figure 13).

Although some carbonation of the existing calcium hydroxides within the concrete can be considered to be occurring, this process can be regarded as a way of introducing carbonation from external sources, causing the calcium carbonate to be formed on the surface, not within the concrete matrix.[35] The depth of full carbonation remains less than 1 mm before and after cleaning.

Conclusions

The conclusions to be drawn from the development of a methodology for the cleaning of the concrete and conservation of the surface finishes of the concrete at Sydney Opera House are two-fold. First, the cause and treatment of surface discoloration has been comprehensively established. Second, through detailed historic understanding of the building as well as sound scientific investigation and a robust philosophical approach, an approach to maintenance has been demonstrated that can be transferred to many situations and materials.

Sydney Opera House has undertaken extensive investigations regarding the causes of discoloration of the folded concrete beams. These have demonstrated that the efflorescence and staining were present, in part, when the slabs were first poured. The variable coloration is due to the

Figure 12 The Utzon Room presentation has been improved by the cleaning and treatment of the concrete without compromising the character of the concrete finish or having to remove early structural patches that help to tell the story of the place. The work sought to reinstate the authentic finish of the exposed concrete, which is such a strong aesthetic element in the building. (Trevor Waters, 2005)

Figure 13 Visual inspection of the soffits confirmed that there were no apparent adverse effects from the cleaning process. (Trevor Waters, 2005)

uneven distribution of calcium hydroxide $Ca(OH)_2$. This is clearly evident in the contemporary photographs taken during construction. Treatment with calcium bicarbonate has successfully evened out the discoloration and brought the beams closer to the architect's original intent. The effect of the cleaning and treatment is to remove sulphates, soiling, and stains, whilst not altering the character of the off-form finish.

The approach taken to developing this treatment is consistent with the unique original design and construction process at Sydney Opera House: that is investigation, experimentation, and implementation. Experimentation on any significant fabric must be undertaken with great care so as not to cause irreparable damage. Through investigation the architect's original intent was established (*Utzon Design Principles*), whilst the conservation management plan provided practical policies for evaluating the heritage significance of the elements to be treated in a current context. Records from the design and construction period further informed the maintenance process. By augmenting documentary records with the collective memory of former workers and supervisors, it was possible to reproduce the techniques employed after 1964. This philosophical approach, adopting considerable investigation before undertaking any work or even experimentation, can be readily applied in many situations and demonstrates the value of understanding the designer's intent before commencing restoration work. This is important when the significance of the place is largely derived from the architectural significance, as is the case at the Sydney Opera House.

This work provides an important record and template for the treatment of the exposed concrete of Sydney Opera House that can be used in the future. By example, this work establishes a documented approach to maintenance and restoration based upon scientific principles and an understanding of the intent and achievements of the original architect and construction workers.

Biography

Susan Macdonald BSc(Arch), Barch, MA(Conservation Studies), RIBA
Susan trained as an architect in Australia before spending ten years in London. A former secretary of DOCOMOMO UK and committee member of Australia ICOMOS, she has a particular interest in the conservation of twentieth-century places and has written three books on this subject. Susan is currently the Assistant Director at the NSW Heritage Office in Australia.

Paul Akhurst BSc(Hons), MSc(Cantab)
Paul has degrees in building management and interdisciplinary design for the built environment, and has previously published on change management. He has spent most of the past twenty years working in the UK and Australian construction and facilities management industries. Paul is currently the Director of Facilities at Sydney Opera House.

Trevor Waters BArch
Trevor is one of the few heritage architects in New South Wales employed as a builder rather than a professional consultant. He has worked in this capacity on a number of important projects in New South Wales including St Mary's Cathedral. Trevor developed the method for cleaning the concrete and supervised the site trials and the execution of treatment of the folded concrete slabs at the Sydney Opera House.

Notes

1 The first paper: Hale, P. and Macdonald, S., 'The Sydney Opera House: An Evolving Icon', *Journal of Architectural Conservation*, Vol 11, No 2 (2005), pp. 7–23.

2 Kerr, J. S., *Sydney Opera House: A Plan for the Conservation of the Sydney Opera House and its Site* (3rd edn.), Sydney Opera House Trust (SOHT), Sydney (2003), p. 36. Available at http://sydneyoperahouse.com/sections/corporate/about_us/pdfs/aboutus_conservationplan2003.pdf accessed 28 September 2005.
This document, which has been formally adopted by the Sydney Opera House Trust, outlines the history and significance of the building and its elements and provides comprehensive policies for its care and conservation.

3 SOHT, *Sydney Opera House Utzon Design Principles*, SOHT, Sydney (2002), p.53. Available at http://sydneyoperahouse.com/sections/corporate/about_us/pdfs/aboutus_utzon_design_principles.pdf accessed 28 September 2005.
This document, written by Jørn Utzon, the building's architect, provides design principles that can be used as a reference for the conservation and evolution of the building. These principles provide a first-hand account of the design drivers or intent that underlay the place.

4 Ibid., p. 78.

5 Ibid., p. 25.

6 Kerr, *op. cit.* (2003), p. 36.

7 SOHT, *op. cit.*, p. 53.

8 Kerr, *op. cit.*, p. 86.

9 Utzon, J., 'Sydney Opera House, Descriptive Narrative, J. Utzon January 1965', unpublished manuscript compiled by M. L. Challenger, SOHT Records at State Archives, Ref. no. SOHDB/24 (Jan., 1965), p. 3.

10 Nicklin, M. S. Associate Partner, Macdonald, Wagner & Priddle, Chartered Engineers (Aust.), 'Sydney Opera House, Summary of Report on Claims Dealing with Increases in Contract Value and Class 'C' Formwork' (Mar. 1962).

11 Levy, A., Resident Engineer, letter to Ove Arup and Partners, London, on the pour of the first pair of beams, No. J6 and J7 (6 Dec. 1960).

12 Roberts, J., 'Memoirs of Jack Joseph Roberts, The Sydney Opera House, 1959 to 1966', unpublished manuscript in the private collection of Mr. J. Roberts, Clerk of Works (Dec. 1994), p. 15.

13 Nielsen, O., letter reporting to Utzon enclosing photographs of the beams soffits of already poured (24 Oct. 1961).

14 Tsoukalas, S., Interview with the author regarding the original techniques for burnishing the in situ concrete beams for the northern foyers of the Opera Theatre and Concert Hall. He has been continuously employed on site for 38 years (2005).

15 SOHT, *op. cit.*, p. 24.

16 Utzon, J., *op. cit.*, p. 17. Quoted in part in SOHT, *op. cit.*, p. 53.

17 Moore, M., GHD Pty Ltd Engineering, *Investigation of Concrete Deterioration & Water Ingress to + 42ft level & Various Locations*, unpublished report prepared for the SOHT, ref no 21/12398/101577/ (Mar., 2004), p. 1.

18 Kerr, *op. cit.*, p. 15.

19 SOHT, *op. cit.*, pp. 78–81.

20 Kerr, J. S., *Reception Hall 'Refurbishment': Draft Comments*, prepared for SOHT at the request of Johnson Pilton Walker Pty Ltd. (3 Sept. 2003).

21 Ibid.

22 Kerr, J. S., *op. cit.*, Policy 41.7, p. 87.

23 Levy, A., Letter to Ove Arup and Partners describing movement and stress cracking in Q beams (10 May 1961).

24 Australian Standard HB84:1996, *Guides to Concrete Repair and Protection*, Section 2.22.3, Fig 2.2.

25 Salome, F., Director, CTI Consultants Pty Ltd., 'Chloride and Sulphate Content of Cement, Opera House', CTI Job: 1443 (27 Feb. 2004), p. 4.

26 Simultaneous differential thermal and thermal-gravimetric analysis (DTA–TGA) was conducted. Of the 31 scrapings collected, 22 samples were of sufficient size to be analysed using a TA Q600 instrument.

27 Paraschiv, V., Senior Development Engineer, Australian Construction Materials, File 627/04 (July 2003).

28 Experimental observations and computer modelling have shown that rainwater or seawater flows also precipitate calcium within the porous system and under certain circumstances the calcium carbonate, with low solubility (0.0013g/100g @ 25°C), may fill the crack, preventing further leaks. (Brodersen, K., *CRACK2 – Modeling Calcium Carbonate Deposition from Bicarbonate Solution in Cracks in Concrete*, Risø National Laboratory, Roskilde, March 2003.) High concentrations of $CaCO_3$ were detected in samples 1 to 5 collected from fractures or cracks with no evidence of moisture. This is in all probability the result of this crack-filling phenomenon.

29 Paraschiv, V., *op. cit.*

30 Nicklin, M. S., 'Sydney Opera House, Stage 1, Arbitration on Folded Slab Formwork', (Nov., 1962) p. 3.

31 Morehead, D., Arup Façade Engineers, 'Folded Concrete Beam Investigation' (Sept. 2003), p. 3.

32 SOHT, *op. cit.*, p. 26.

33 Bartholin. E., *Experimental Crystalli Islandici Disiaclastici* (1669).

34 Paraschiv, V., *op. cit.*

35 The introduced carbon dioxide could be increasing the depth of carbonation of the concrete, reducing its alkalinity and making the steel reinforcement more vulnerable to corrosion. In practice the calcium carbonate formation appears to be superficial only and the depth to which the alkalinity has dropped below 9.2 pH within the concrete matrix is unchanged.

Saint John the Divine

Techniques to Assess Fire Soil

Claudia Kavenagh and Christopher John Gembinski

Abstract

The Cathedral Church of Saint John the Divine is the third largest church structure in the world. It is located on the upper west side of Manhattan in New York City. A fire in December of 2001 badly damaged the stonework of the unfinished north transept and spread thick black smoke through the entire building. After the fire, a comprehensive investigation was undertaken to identify the unique characteristics of the fire soil and then to use that information to determine the extent to which these materials had deposited on surfaces in the building. A direct correlation was made between the materials of construction that burned during the fire and the particles present in samples of soiling removed from throughout the Cathedral. It was therefore possible to establish that the fire soil had penetrated to all areas of the Cathedral and that all surfaces required cleaning. This paper describes the techniques used to carry out the investigation.

Introduction

A fire in December of 2001 at the Cathedral Church of Saint John the Divine resulted in significant damage to architectural stonework in and adjacent to the Cathedral's north transept and caused smoke to fill the entire Cathedral interior (Figure 1). In the days following the fire, assessment was made of the effects of the fire on the exterior and interior architectural elements of the Cathedral building. An important initial component of the assessment was to characterize the soiling left behind after the smoke cleared and to catalogue the surfaces on which fire soil was deposited.

The initial assessment indicated that the soiling from the fire had deposited on every surface throughout the building and that all surfaces

Figure 1 A view of the north transept during the fire, taken from an exterior parking area. (Michael Schwartz/New York Daily News)

would have to be cleaned. Because of the enormous cost associated with such an undertaking, the insurance claim for the damage caused by the fire had to include comprehensive and irrefutable documentation of these conclusions. It was therefore essential first to prove that the soiling from the fire had affected the entire interior and then to determine an appropriate means for cleaning the materials, so that the cost of this work could be established.

Surfaces exposed to smoke conditions and contaminated by-products of combustion are typically darkened or, in extreme cases, blackened. The composition of the particulate contamination depends on the articles that burned. Smoke and other by-products of the fire that were transported and subsequently deposited on surfaces in the Cathedral are generically referred to in this paper as fire soil.

History and description

The Cathedral Church of Saint John the Divine is the third largest church structure in the world and the largest cathedral in the French Gothic style.[1,2] It is located on the upper west side of Manhattan in New York City. The Cathedral was constructed in phases beginning in 1892. Architects George Lewis Heins and Christopher Grant Lafarge prepared the first designs for the structure in the Romanesque style. The initial phase of construction lasted from 1892 until 1911. In 1916, Ralph A. Cram took over as chief architect and modified the earlier Romanesque design to incorporate the French Gothic style that is visible today. Construction during this phase continued until the start of World War II. Work then occurred sporadically over the next few decades. The most recent phase of construction began in 1979 and ended in the 1980s. Although in use since its initial construction period, the Cathedral is still an unfinished structure. Most notably, the west (main) elevation and the two towers flanking the main entrance are only partially completed. On the interior, primarily in and adjacent to the crossing, cladding stone was never installed and the inner masonry wythes are exposed.

The Cathedral is a massive construction spanning a reported 183 metres from the narthex to the back of the apse and 38 metres from the floor to the vaulted ceiling. In addition to the architectural materials, the Cathedral has an extensive art collection. The much-loved Cathedral is constantly active and services are held in the building several times daily. The building is used by a school on the grounds and by theatrical companies; there are numerous concerts and other events throughout the year. Each day, hundreds of neighbourhood residents and tourists come to visit the Cathedral.

The primary material of construction on the Cathedral exterior is Mohegan granite, with some limestone trim elements. A fossiliferous calcitic limestone, most likely an Indiana limestone, and a smooth sawn granite, clad the nave and choir interior walls. Interior wythes of masonry are visible in the crossing and ambulatories, where the final cladding stone has not yet been installed. On the majority of unfinished walls, a rough cut

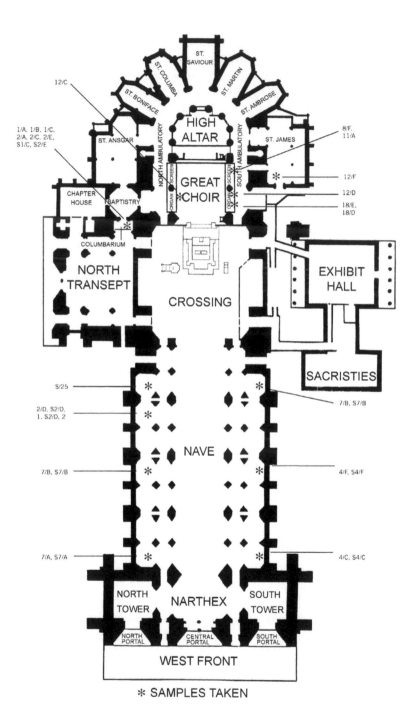

Figure 2 Plan of the Cathedral showing representative sample locations. Samples were collected at a variety of locations, close to and far from the north transept, and at different heights at each location.

granite is exposed. Common brick was also used for interior masonry wythes, though less typically than granite, and is exposed at a few locations in the crossing and ambulatories. The temporary north and south walls of the crossing are poured-in-place concrete. The dome of the crossing and the vaults of the ambulatory are Guastavino tiles. The ceiling vaults in the nave and choir are finished with a sound-absorbing type of Guastavino tile known as Akoustilith.

The fire

The fire that occurred in December of 2001 began in a small room in the northeast corner of the north transept. A temporary roof had been built on the unfinished north transept to allow for an interior use and comprised wood framing and sheathing, foam insulation, and bitumen-based built-up roofing. Since the 1980s, the ground level of the north transept had been used as a gift shop and this area was stocked with merchandise for the holiday season at the time of the fire. The fire began as a slow burning electrical fire which smouldered throughout the night and caught flame early the next morning.[3]

When the fire reached the temporary roof, it spread rapidly. However, since this roof structure was the only wood framing in the Cathedral, the fire did not spread rapidly to other areas of the building.

As the fire burned, it extended beyond the confines of the north transept at three general locations:

- onto the exterior north wall of the Cathedral, above the roof of the north transept;
- through the temporary concrete north wall of the crossing; and
- through a temporary wall construction of wood stud and gypsum dry-wall between the north transept and the columbarium.

The primary materials that burned were the roof and the merchandise in the gift shop. Two seventeenth-century tapestries that had been hanging on the north wall of the crossing were badly damaged by the fire. Eyewitnesses report that thick black smoke from the fire travelled throughout the entire interior of the Cathedral. The New York City Fire Department brought the fire under control within several hours (Figure 3).

The initial response

At the time of the fire, the Cathedral staff were able to respond immediately. Since the fire occurred in the early morning, some of the Cathedral staff were already at work. The Cathedral staff mobilized rapidly and acted as escorts for the firefighters. The escorts, who had good knowledge of the

Figure 3 A view of the interior of the north transept after the fire. Charred wood beams and partially melted bituminous materials from the roof structure are in piles on the floor. (Michael Schwartz/New York Daily News)

layout of the building, guided the Fire Department through the interior and assisted them by opening locked doors. The Cathedral was therefore able to avoid destruction of many valuable interior elements that would have otherwise been lost or damaged.

The smoke that filled the entire Cathedral greatly reduced visibility. This created difficulties for the firefighters, who were not familiar with the floor plan of this substantial building. It was suggested to the Cathedral staff that some of the stained glass windows be broken to allow the smoke to escape the nave. The Cathedral staff made the immediate decision not to break any of the stained glass windows with the knowledge that they would have to address the increased deposition of fire soil on interior surfaces that would occur.

After the fire was extinguished, the Cathedral wanted to re-open for holiday events and services. Because the fire occurred shortly before Christmas and large crowds were expected, it was decided that some measures were necessary to protect parishioners and tourists from contact with the heavy build up of particulate matter from the fire soil. The Cathedral staff mobilized a work crew that removed loose particulate matter from wall surfaces in the main interior of the Cathedral, from the floor up to a height of approximately 2.5 metres above the floor. Removal of the soiling was accomplished by wiping the surfaces with dry urethane sponges, similar to art sponges. The objective was to remove as much loose particulate matter as possible from the walls with a technique that would not damage the surfaces or leave a residue. The Cathedral successfully opened for Christmas 2001 services and has remained open to the public ever since.

The measured response

Initial assessment

The minimal intervention of limited sponge cleaning allowed for a rapid re-opening of the building because the cleaning eliminated the risk that the public would come in contact with fire soil. It was then possible to research a logical long-term response and develop a long-range plan to address the damage and soiling caused by the fire.

The primary exterior materials that were affected by the fire were the smooth-sawn limestone, carved limestone, rough-cut granite, and poured-in-place concrete. The limestone and granite masonry in the north transept suffered severe spalling related to the heat of the fire and the rapid cooling caused by the use of water to put out the fire. Prevalent conditions included cracked, shattered, and spalled stone and thick accretions of a tar-like substance and other soiling. As the fire cooled and water infiltrated the now exterior stone walls, spalls continued to develop for days after the fire. Smaller losses occurred for some months (Figure 4).

The primary interior materials affected by the fire were the smooth-sawn limestone cladding and carved limestone, rough-cut and smooth granite, architectural woodwork, and the Akoustilith tile of the ceiling vaults. Smoke from the fire also travelled into the non-public spaces of the building, such as between the vaulted ceilings and the roofs, and into the circular stairways of the great choir and some of the access tower passages. The stairways and tower passages acted as chimneys, as smoke was drawn up through these areas, and large amounts of fire soil were deposited on the surfaces within them.

Additional materials in the interior affected by the fire included:

- limestone with polychrome coatings in the baptistery;
- decorative finish materials located primarily in the chapels, such as polished marble base boards and other trim elements, face brick with surface texture, ornamental metalwork, glazed ceramic tile, and mosaic tile;
- the monumental polished granite columns in the choir;
- ornamental woodwork with stain and clear coat finish lining the inner walls of the ambulatory;
- various pieces of artwork throughout the building;
- and the great organ and its related elements.

A series of objectives were developed to address the fire damage at the Cathedral systematically. The list of main objectives included:

- determine the extent to which the fire soil infiltrated the building;
- determine the appropriate methods for the removal of the fire soil;
- evaluate the damage to and potential restoration of the stone in the north transept;
- develop a plan for cleaning the entire interior of the building.

For purposes of this paper and in the context of this publication, the following discussion focuses on conditions and treatments related to soiling associated with the fire. Damage to the building, and repair and restoration work unrelated to cleaning treatments, are not further discussed here.

The fire soil darkened all surfaces to some degree; some materials were completely black. In addition to the north transept itself, the heaviest concentrations of soiling were directly adjacent to this space in two general locations: the baptistery and the columbarium. However, in other areas, it was not possible to macroscopically verify the presence of fire soil. The 109-year-old Cathedral had never been cleaned and, prior to the fire, interior surfaces had varying amounts of soiling on them. No comprehensive documentation of pre-fire soil conditions existed, so that it was not possible to prove by simple visual comparison that the coloration of surfaces was altered by the fire. In order to present a strong case for the insurance claim, it was necessary to provide clear evidence of the presence of fire soil.

Figure 4 The limestone in the north transept exhibits spalls from the stresses caused by the heat from the fire followed by sudden cooling when the water used to extinguish the fire hit the stone. The stone was blackened by the deposition of smoke, which was heavily laden with bitumen.

Sampling at the source of the fire

In order to determine on which surfaces in the Cathedral's interior fire soil was deposited during the fire, it was first necessary to identify and characterize fire and smoke deposits from known sources and then compare the identified fire soil with samples removed from throughout the interior of the Cathedral.

Samples of burned materials from the north transept were collected for analysis and placed in sealed plastic bags. These samples were collected from representative areas that included a partially burnt wood beam and several burnt modern roofing materials from the temporary roof and charred limestone spalls. In addition, three small objects (two ceramic figurines and a metal tea tin) that had been exhibited for sale in enclosed display cases at the time of the fire were removed and placed in sealed plastic bags. It was assumed that these items would have been relatively clean and free from soiling before the start of the fire. Accordingly, the soiling found on these items could unquestionably be documented as fire soil and therefore could serve as a means of comparison with soiling samples taken at other locations.

Visual microscopy was the primary means of comparing confirmed fire soil with unknown soiling samples removed from various locations throughout the building. Morphological characteristics of the particles found in the soiling, including colour, texture, shape, and size, were primary identifiers. In addition, semi-quantification of the amount of fire soil particles deposited in specific areas was attempted using visual microscopy.

First, visual characterization was performed of the typical burnt materials of construction in the north transept: pieces of charred wood, melted modern roofing materials, insulation and stone. Then, the soiling on the two ceramic figurines and the metal tea tin was examined and compared with the burned materials of construction.

Visual characterization was performed using a Nikon SMZ-U stereobinocular microscope with fibre-optic light source and a magnification up to 75X, as well as a Zeiss SV-11 stereobinocular microscope with fibre-optic illumination and a magnification up to 125X. Under magnification, small particles of burned construction materials and charred wood fibres were observed in the soiling found on the two ceramic figurines and the metal tea tin. Soil particle sizes ranged from very large fragments (up to 1 mm in length and visible to the naked eye) to very small fragments visible only under the highest magnification (125X). The particles were primarily rectangular or oblong fragments resembling thick splinters of charred wood. In addition, a black glassy substance visually resembling bitumen or tar was observed in varying shapes and sizes including perfect spheres, fragmented spheres and attenuated globules. The microscopic particles resembled the large globules of fused material found on the samples of building materials. Various other inorganic materials observed in the soil included: stone and mortar fragments; sand particles; and melted modern construction materials such as roofing felt and insulation (Figures 5a, 5b, 5c and 5d).

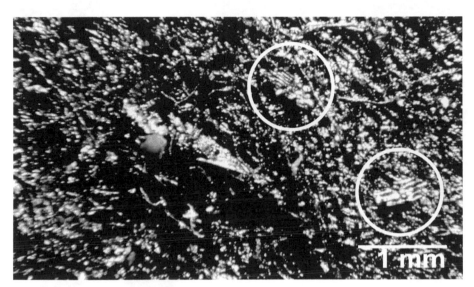

Figure 5a Particles from charred wood beams (in white circles) found on fire damaged roofing materials recovered from the north transept (25X).

Figure 5b Charred wood particle isolated from a sample taken from the north transept represents the typical shape and colour of the smaller particles found in samples taken throughout the Cathedral (62.5X).

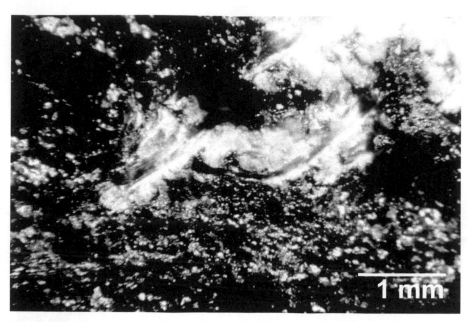

Figure 5c Burned bituminous roofing material on stone recovered from the north transept (25X).

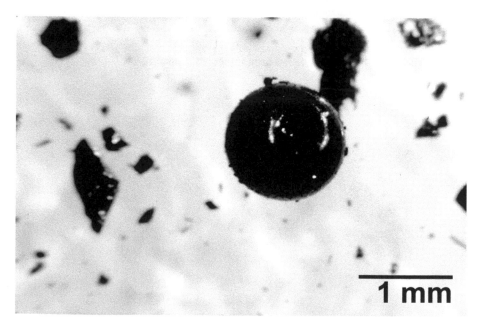

Figure 5d Large bitumen sample taken from the north transept represents the typical shape and colour of the smaller particles found in samples taken throughout the Cathedral (25X).

Through this microscopic examination, the source of the particles found in the soiling on the figurines and the tea tin – soil that was known to be a result of the fire – could be directly ascribed to the materials of construction that burned during the fire. In this way, the typical morphology of the fire soil was established. The characteristics identified in these observations became the primary identifiers for fire soil.

Interior sampling method

Once the characteristics of the fire soil were determined, the next step in the process was to confirm the deposition of fire soil on the interior surfaces. Accordingly, 144 soiling samples were systematically collected from the architectural surfaces. Samples were obtained from alternate bays in the nave at various heights from the floor to the ceiling vaults as well as in the ambulatory, chapels, columbarium, baptistery, spiral staircases, and roof access ladder wells. In addition, several samples were collected from behind a painting that was hanging on the wall of the Chapel of Saint Saviour at the time of the fire. It was hypothesized that the samples removed from behind the painting would represent the typical atmospheric soiling found on the interior of the Cathedral prior to the fire.

A two-part sampling method was developed for general qualitative analysis and semi-quantitative analysis. The first method used double-sided tape adhered to paper index cards. The tape and cards were pressed against the soiled stone surface, lightly rubbed, and then peeled back. The fire soil adhered to the double-sided tape (Figure 6). The cards were labelled and then stored in plastic cases manufactured for storage of compact discs. This method proved beneficial in identifying, cataloguing, protecting, and storing the samples. The tape method allowed for characterization of the fire soil under microscopic examination, and provided a semi-quantitative analysis of the concentration of the fire soil in specific areas.

Figure 6 Double-sided tape was adhered to paper index cards and then pressed against the stone surface to remove samples of fire soil.

The second sample method used sterile rayon/cellulose gauze pads wiped across the soiled surfaces of the areas adjacent to the tape samples. The pads removed bulk samples of the fire soil for further analysis. They also served as a back-up method for areas where no fire soil appeared to attach to the tape.

In isolated areas, such as on polished marble surfaces, it was difficult to ascertain if any fire soil was collected by the above methods. In these areas samples were removed using a vacuum with a soft clean brush attachment and a fresh bag for each location. Following collection, the vacuum bags were placed in sealed plastic bags.

Characterization of samples

A systematic microscopic examination was conducted of the samples removed with tape from throughout the Cathedral. Observations showed that the charred wood fibres and the glassy bitumen globules were present in all samples, from the ones closest to the fire removed at the columbarium to the ones farthest away at the narthex. The charred wood fibres and the glassy bitumen globules were even found on the surfaces that had been dry wiped with a sponge immediately following the fire. Generally, the quantity of particles and particle size distribution varied depending on the distance from the fire and whether or not the surface had been partially cleaned; smaller quantity and size were typically found in the samples from locations farthest away from the fire and where the wall had been wiped with the dry sponges. However, the characteristics of the soil particles were virtually identical on every sample taken (Figures 7a, 7b, 7c, and 7d).

The only exception to this finding was the samples taken from behind the painting in the Chapel of Saint Saviour. Identified fire soil was not found in these samples. Lack of fire soil in this protected area allowed for a clear distinction between fire soil and the general, older atmospheric soiling present on the stone. Combined with evidence of fire soil on the unprotected stonework near the painting, lack of fire soil behind the painting offered further evidence of the spread of the fire soil.

No stone or sand particles and none of the more fibrous melted materials were observed on the samples removed with tape from the interior masonry surfaces. This suggests that the heaviest particles in the smoke from the fire fell out of the air relatively quickly and only the charred wood fibres and the bituminous particles travelled to the interior of the Cathedral. The bitumen particles are heavier than the charred wood fibres; the highest concentrations of bitumen particles were found closest to the fire and were most prevalent in samples removed from the columbarium. A high concentration of charred wood fibres, with relatively few particles

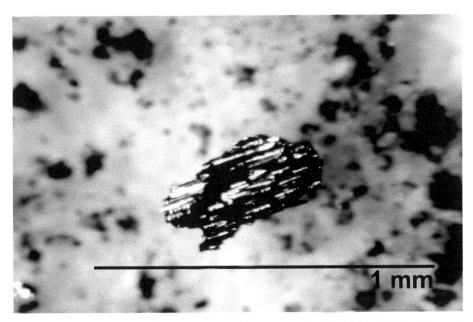

Figure 7a Sample removed from a bay in the nave at a height of 6 m (100X). Charred wood particle matches the morphology of those found in the north transept (as seen in Figures 5a and 5b), but is much smaller in size.

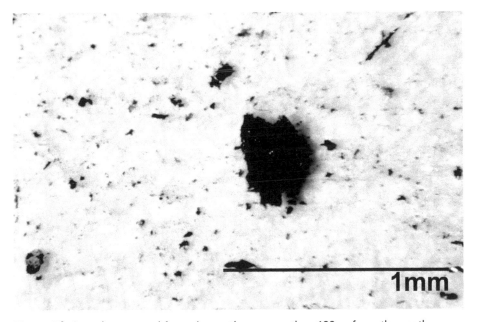

Figure 7b Sample removed from the narthex, more than 122 m from the north transept (62.5X). Charred wood particles reduce in size farther from the north transept.

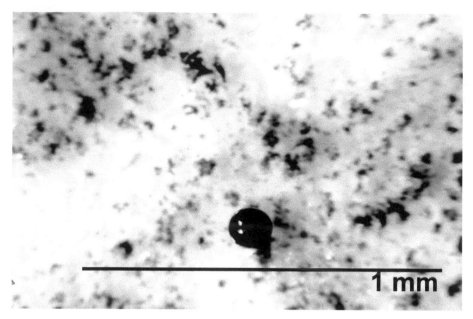

Figure 7c Sample removed from a bay in the nave at a height of 6 m (62.5X). Bitumen globule is much smaller in size than the one seen in Figure 5d but exactly matches in shape and colour.

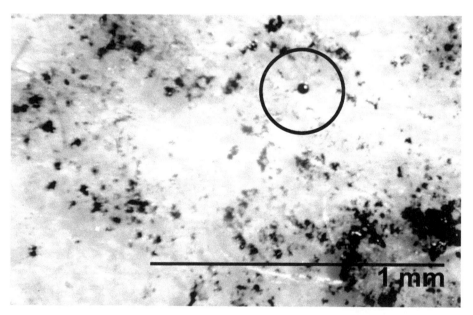

Figure 7d Sample removed from the narthex, more than 122 m from the north transept (80X). Small bitumen globule (in black circle) matches the shape and colour of those seen in both Figures 5d and 7c.

of bitumen present, was found at sampling locations farthest from the fire. In addition, very few bitumen particles were found in the samples removed from the ceilings.

Microscopic analysis of the samples removed with tape provided clear evidence of the prevalence of the fire soil in most cases, Therefore, with the exception of the Akoustilith ceiling tile, where it was not possible to remove a sample using tape, the bulk samples taken with sterile cotton were not analysed.

The analysis provided sufficient documentation to successfully file an insurance claim proving that fire soil penetrated to all areas of the building and that, consequently, there was a need to clean all interior surfaces. This was a crucial step in the planning process. With the knowledge that insurance monies would be available to clean the entire building, the Cathedral staff were able to move forward confidently with the development of a programme for comprehensive interior cleaning.

Cleaning test program

The next step in the process was the development of a comprehensive cleaning test programme for all interior materials. The large majority of interior surfaces are clad with either limestone or granite; there are over one million square feet of limestone and granite cladding on the interior of the Cathedral. Therefore, these stones became the primary focus of the cleaning test program. A cleaning test program was also developed for all other interior cladding materials, however, it will not be discussed further in this paper.

The fire soil on the granite and limestone was divided into two categories: lightly and moderately soiled stone (in areas more removed from the source of the fire); and heavily soiled stone (at and directly adjacent to the source of the fire). The analysis of fire soil had shown that the primary components of the fire soil were different close to and far from the source of the fire. The heavy bitumen particles fell out of the smoke more quickly and deposited on surfaces close to the fire. The charred wood and other light particulate matter travelled farther. The removal of the bitumen requires very different products and techniques from the remainder of the soil types.

Since an analysis of the soil on the walls where the initial cleaning had been performed using sponges showed that fire soil still remained on these surfaces, it was known that mild mechanical methods could not completely remove the fire soil. The two basic categories of cleaning methods for architectural surfaces of significant size are wet chemical cleaning and abrasive cleaning. Neither of these methods permits selective removal of fire soil from masonry. An attempt to partially remove soiling from the masonry carries the inherent risk of producing a surface that is not uniform in appearance, which would not be an improvement on the appearance of the building prior to the fire. It was therefore deemed necessary to develop cleaning methods that would remove both the fire soil, and the general soiling and stains that had developed over the course of time.

Another important consideration was that all work would be done on the interior of an occupied building. Although the Cathedral was large enough to permit work to proceed in one protected area while public activities continued in other areas, there was still a risk of the transport of cleaning agents, the migration of odours of cleaning agents, and the travel of noise created by the cleaning method. In addition, the control of water used to clean or rinse masonry is difficult on a building interior.

A third consideration was that, given the enormous size of the building and the consequent logistics of assembling equipment and performing any one cleaning method, it was desirable to develop a single method that would successfully clean both the limestone and the granite.

The generic types of chemical cleaning agents and methodologies that are appropriate for use on limestone and granite have been thoroughly addressed by others.[4] Limestone is a relatively soft calcareous stone. Granite is a much harder, essentially siliceous stone. The different mineral components of the two stones react differently when subjected to different chemical cleaning agents and, during abrasive cleaning or water rinsing, will safely withstand very different levels of pressure. Therefore, although similar cleaning tests were performed on both stones, the tests were tailored to ensure appropriate methods for the two types of stone.

A series of small (approximately 20 cm × 20 cm) wet chemical cleaning tests were conducted on both the limestone and the granite. A range of chemical agents were employed, including acids, bases, bases and acids applied in sequence, solvent, and detergent cleaners. The same general procedure was used for each test: the surface was pre-wet (if required by the type of chemical used), then the chemical was applied and kept on the surface for a pre-determined period of time, the surface was then scrubbed with a soft bristled brush, and it was finally rinsed with pressurized water. Very low pressure (less than 100 psi) was used wherever possible. When necessary to ensure complete removal of chemical, a 500–700 psi water rinse was employed (Figure 8).

Several different wet chemical cleaning methods successfully cleaned the lightly and moderately soiled surfaces. None of the tests were successful in removing the heavy bitumen build-up. A number of issues indicated that wet chemical cleaning would not be appropriate. Strong acids and bases have inherent risks in an interior occupied space and some chemicals have pungent odours that are unacceptable. Other chemicals release gases that

Figure 8 Tests were conducted to determine the effectiveness of wet chemical cleaning for the removal of the fire soil and pre-fire general soiling.

could have deleterious effects on interior materials such as glass and wood. All chemical cleaning agents require a thorough rinsing. Given the extent of surfaces to be cleaned, there would be an enormous amount of water to contain, collect, and dispose of. Even with thorough rinsing, chemical products can introduce the risk of leaving residues in the stone, where they can cause later damage. Use of copious amounts of water also introduced the risk that older soiling within the stone might be displaced by the cleaning action. Both organic materials and salts that had infiltrated the stone over the years – primarily from leaks in exterior walls – might be mobilized, creating additional stain issues and, more importantly, issues related to salt crystallization within the stone units.[5]

Low pressure abrasive methods were also tested, using a variety of abrasive media including dry ice, particles of sponge, particles of sponge with abrasives embedded in them, and mineral powders.[6] At very low pressures (less than 50 psi), both particles of sponge with abrasives embedded in them and mineral powders appeared to clean the lightly and moderately soiled stone without damaging the substrate. Dry ice caused damage to the stone at any pressure. None of the tests were successful in removing the heavy bitumen build-up. In order to confirm that the abrasive methods had not altered the substrate of successfully cleaned stone, 12 mm diameter cores were removed for analysis. Each core was located so that half the core was of the cleaned surface and the other half had not been abrasively cleaned. Each removed core was then split in half lengthwise and the surface was viewed in cross section under magnification. In this way, the surface topography of cleaned and uncleaned stone could be compared. Results showed that the surface of the stone had not been altered by the cleaning method. However, a number of issues indicated that abrasive cleaning would not be appropriate. The containment of abrasive media was of particular concern; not only would full containment of the work area be very costly, but any break in the protection would result in rapid spread of media to other areas. Importantly, the success of an abrasive system is highly dependent on the skill of the worker and his ability to achieve consistent results. Variations in working distance from the surface to be cleaned, in the amount of time spent cleaning each area, and even in the specific arm movement used can result in significant visible variations on the cleaned surface. Given the quantity of large flat surfaces to be cleaned, this was deemed a high risk factor. Finally, with proprietary abrasive systems it is not possible to follow the typical process in which a number of masonry cleaning contractors submit pricing for a project, and then the most competitive bid is awarded the job.

A series of tests were also conducted using a liquid latex with and without chemical additives as the cleaning agent.[7] All surfaces were vacuumed prior to application. The latex was applied to the surface using spray equipment. It remained on the surface until cured, at which time it was peeled off in sheets. The masonry was scrubbed with a soft bristled brush and then wiped with sponges dampened with water to remove residue. The latex with a chemical additive successfully cleaned the light and moderately soiled stone, but did not clean the heavily soiled stone. It was theorized that the latex would keep the chemical additive at or near the surface of the masonry, making it less likely that the salt component of the chemical would be absorbed into the stone and minimizing the amount of water needed to rinse the stone. Testing confirmed that this theory was correct.[8] The spray application was useful in that it would be possible to cover large surfaces in a short period of time. However, there was still concern that, despite protective measures, overspray could migrate to other areas in the building. The risk was felt to be much lower than for abrasive systems because the latex is heavy and tends to fall out of the air close to the expulsion point.

The positive and negative aspects of each cleaning system were assessed. The decision was made to clean the lightly and moderately soiled granite and limestone using the latex system. Because no method was found to clean the most heavily soiled stone, consideration is being given to replacing this stone in some cases and leaving it in its soiled state in others. Attics and some other secondary, non-public spaces will be vacuumed only, to ensure that no loose fire soil can migrate to cleaned spaces.

Cleaning began in February 2005. The work is being performed in phases so that at all times some areas of the building are open to the public. It is anticipated that all work will be complete within three years. The work is being performed by a masonry cleaning contractor, and monitored by a team of architectural conservators.

Conclusions

A direct correlation was made between the materials of construction that burned during the fire and the particles present in the samples of soiling removed from throughout the Cathedral. It was therefore possible to establish that the soil from the fire had penetrated to all areas of the Cathedral. Evidence of fire-related soil was found from the narthex to the ambulatory and from the lowest sections of the walls to the ceiling vaults. Evidence was also found in non-public areas, such as enclosed stair halls, walkways above the chapels, and the interstitial space between the ceiling vaults and the roof.

In almost every sample examined, charred wood particles and bitumen globules were present. The particles had the same morphologies, matching in colour, texture and shape. The one major differentiating characteristic was their size; variation in size had a direct correlation to the distance of the sample from the fire.

The immediate organization and the response of the Cathedral staff working in cooperation with the New York City Fire Department saved priceless materials and artwork in the building. The decision to remove a minimal amount of the loose particulate matter initially, using a method that did not adversely affect the existing historic materials, was a solution that did not interfere with the creation of a measured response. While the Cathedral continued to function, the project team was provided the time and opportunity to develop a testing matrix and use analytical techniques to characterize the fire soil and confirm that the fire was the source of the soiling found throughout the building. The systematic documentation contributed to the successful settlement of an insurance claim.

The information then assisted in the development of a comprehensive cleaning test program for all interior materials. The testing program resulted in the development of specific cleaning techniques for removal of fire soil. Cleaning of the Cathedral interior is currently in progress.

Biography

Claudia Kavenagh
Claudia Kavenagh received her MS in historic preservation from Columbia University's Graduate School of Architecture, Preservation, and Planning. She is director of the New York offices of Building Conservation Associates Inc (BCA), where her work combines the disciplines of historic preservation and materials conservation for the restoration of buildings and monuments.

Christopher John Gembinski
Christopher Gembinski has been with Building Conservation Associates Inc since 1998. Through his work on numerous large-scale projects, he is versed in restoration repairs for a wide range of historic building materials. As an experienced conservator, he has performed laboratory analyses using analytical techniques including visual microscopy, x-ray diffraction, ultrasonic testing and Fourier transfer infrared spectroscopy to characterize historic architectural materials. He has also worked for the contracting firm Archa Technology, Ltd. as the superintendent and project manager. He is a graduate of the University of New Hampshire and received an MS in preservation from the University of Pennsylvania.

Acknowledgements

Materials analysis was performed by George Wheeler, Director of Conservation in the Department of Historic Preservation at Columbia University and research scientist in the Department of Scientific Research at the Metropolitan Museum of Art. BCA's project team for the field and laboratory work also included Kevin Daly and Richard Pounds.

Notes

1 Hall, E. H., *A Guide to the Cathedral Church of Saint John the Divine in the City of New York*, The Cathedral Church of Saint John the Divine, New York (1965).
2 Dolkart, A. S., *Morningside Height, A history of its architecture and development*, Columbia University Press, New York (1998).
3 LeMire, J., Ortega, R., and Siemaszko, C., 'Five-Alarmer at Cathedral', *New York Daily News*, 19 December 2001, p. 3.
4 There are numerous published texts on masonry cleaning. One comprehensive resource is Ashurst, N., *Cleaning Historic Buildings*: Vol. 1. *Substrates, soiling and investigations*; Vol. 2. *Cleaning materials and processes*, Donhead Publishing, London (1994).
5 Winkler, E. M., *Stone: properties, durability in man's environment*, Springer-Verlag, New York (1975).
6 Proprietary and non-proprietary low pressure abrasive systems were tested. Proprietary systems included Façade Gommage® and Sponge-Jet.
7 Proprietary product used was Arte Mundit® with an EDTA additive, manufactured by FTB Restoration.
8 Kavenagh, C. and Wheeler, G., 'Evaluation of Cleaning Methods for the Exterior Brick at the Brooklyn Historical Society', *Journal of the American Institute for Conservation*, Vol. 42, 2003, pp. 97–112.

St Paul's Cathedral

Poultice Cleaning of the Interior

Martin Stancliffe, Inge De Witte and Eddy De Witte

Abstract

The construction of St Paul's Cathedral, from the laying of its foundation stone in 1675 to its declared completion in 1710, was the product of the vision and determination of Sir Christopher Wren. But the intervening years have seen the interior paintwork removed, leaving the stonework to become increasingly dirty. A superficial cleaning programme was carried out in the 1930s, but it failed to address the essential problem of the stained and soiled stonework. By the 1990s, concern arose about the dirty condition of the interior. None of the cleaning systems developed in the past twenty years is systematically used for large scale cleaning of interiors of historic buildings. A recent development in interior cleaning is the introduction of peelable poultices based on a specially formulated natural latex dispersion. This paper looks specifically at the conservation methods used for cleaning the interior of St Paul's Cathedral. The paper also focuses on the technical development of the basic poultice, Arte Mundit® type I, a custom formulated aqueous dispersion of natural rubber which is designed for application in historic buildings. The results of cleaning studies using this system on the interior of St Paul's are discussed in detail.

Introduction

When he visited St Paul's Cathedral in June 1710, the German visitor Conrad von Uffenbach was overwhelmed by the 'extraordinary lightness' of the interior of the recently completed Cathedral. But by the early decades of the late twentieth century, the stonework had become a dark and mottled brown, and the whole interior was both dirty and gloomy.

The cleaning of so large and so important a building was clearly going to require a substantial amount of skill, time, and money; and a project of this nature cannot easily be justified when more pressing work is required. But by 1994 other important outstanding issues had begun to be addressed, and it became clear that the opportunity to consider such an undertaking might be grasped (Figure 1).

The interior cleaning of St Paul's Cathedral was completed in May of 2005 after four years on site, but the process actually began more than ten years before, in 1994. It is not only the execution of a work of this nature which is challenging, but also the preparation that lays the foundation for the success of the undertaking.

In selecting a cleaning process for this project, it was noted early on that a variety of techniques has been and still are being developed for the cleaning of historic façades. Although the evolution of these techniques is influenced by geographic traditions as well as the philosophical approaches of local decision makers, one can observe a chronology in the evolution of cleaning techniques all over Europe. It begins with dry and wet sand blasting, followed by the use of chemicals, high and low pressure jets, Façade Gommage® (the proprietary system provided by Thomann-Hanry) and rotation whirl systems, and finally laser cleaning. These are all techniques which have been used consistently over the last two decades. None of these systems is used for large scale cleaning of interiors of historic buildings: either too much water has to be used, which causes damage to the interior (such as parquet flooring, furniture, etc.), or too much dust is generated. When interior cleaning with such techniques has been accomplished, the removal of all furniture and careful protection with plastic sheets, including construction of nearly airtight tents in some cases, has increased the cost to a sometimes unacceptable level. Laser cleaning is increasingly used for the cleaning of statues, but for large flat surfaces the cost-effectiveness is too low. In some cases, it also provokes noticeable colour changes due to interaction of the laser with coloured components, such as glauconite.

Historical background

A proper understanding of the existing conditions was an essential starting point. The interior of St Paul's Cathedral appeared very different in the 1990s from how it was when von Uffenbach visited. Shortly before its completion, Wren had arranged for all of the interior stonework to be painted '3 times in oyle with culler'. This light stone-coloured paint coating was a finish Wren normally gave to the interiors of his city churches: he believed that the application of oil hardened and protected the stone surface. But there may have been other reasons too. The coating may have been intended as a preparation for a painted decorative scheme on the lines of the Painted Hall at Greenwich; or to cover the effects of the new stone

Figure 1 Completed interior cleaning of St Paul's Cathedral. View from the West Triforium Gallery. (Angelo Hornak)

inserted where the supports for the dome had cracked; or simply to unite and brighten an interior already dirtied by exposure to the coal-smoke polluted atmosphere of London. Whatever the original intention, all of this paintwork had been stripped off between 1872 and 1874 with an enormous expenditure of effort, caustic paste, and abrasion, perhaps in pursuit of the Victorian ideal of purity of material. The resulting finish was far from satisfactory: oil had penetrated deep into the pores of the stone, trapping dirt from coal combustion and gas lighting systems. The surfaces were further damaged during major structural works in the 1920s. At this time a considerable amount of new stone was inserted in the piers of the dome, and these piers were then grouted. Liquid grout spilled extensively down the stonework. This again was abraded off, and comprehensive cleaning was carried out in the 1930s, using water and stiff brushes. This process also sought to remove the extensive salt encrustations believed to have resulted from the caustic paste used in the 1870s. But this programme of cleaning failed to get to grips with the real essence of the problem – the dark, mottled appearance of the stonework, and the resulting dirty, gloomy character of the interior spaces – and was not really successful.

Since the completion of Wren's Cathedral, many other changes had taken place within the interior. Wren's original layout had been drastically changed in the 1860s by the removal of his organ screen; and between 1890 and 1907 the east end of the Cathedral was covered by mosaics in the Byzantine style, designed by Sir William Richmond. Related to this work were the remains of painted and gilded surface decorations dating from a number of different periods in the second half of the nineteenth century, and largely, but not entirely, removed in the 1920s.

In the light of the remnants of painting and gilding, it was agreed at the outset that the stonework could not simply be repainted, as the work carried out since the mid-nineteenth century had fundamentally changed the interior. In addition, to repaint the stonework would create a future maintenance problem, as well as result in a finish that might seem inappropriate to modern eyes.

Initial studies

In addition to the understanding of the historical context as briefly summarized above, and as much further detail was required as the project developed, there were substantial technical questions to face from the outset. Was the nature of the surface soiling fully understood? Was it possible to devise a project that could deliver a technically acceptable result? Could this be done at a reasonable cost, without unacceptable disturbance to the life and work of a busy cathedral, and within a reasonable time-scale? Were suitable contractors available with the required capacity and skills?

Between 1994 and 1998 much preliminary work was done to provide initial answers to these questions. Deborah Carthy, an experienced stonework conservator, evaluated a range of possible cleaning techniques. There were also investigations into the remaining paint traces (fragments of which could be found in various places within the crevices of the stonework) and into documentary evidence for the original paint. Detailed research was carried out into the history of the decorations of the interior; and there were extensive discussions with the Cathedral's Fabric Advisory Committee.

The tests carried out by Deborah Carthy assessed a range of available cleaning techniques. These covered:

- Low pressure steam (4–5 psi)
- De-ionised water and white spirit in equal parts with Synperonic N (a surfactant)
- Vulpex soap in both water and white spirit
- Gravity fed air abrasive (at less than 40 psi) using JBlast Finesse
- Mora pack including ethylenediaminetetraacetic acid (EDTA), a chelating agent
- Laser cleaning

It is noted that the advanced laser cleaning system was considered in concept only. Upon close consideration, it would have been a costly system to utilize given the amount of time required to clean the large areas of the interior of the Cathedral.

Of these systems, only the air abrasive and Mora packs had any real cleaning effect. The micro air abrasives proved highly effective; but the costs involved with this technique would have been prohibitive, because of the protection which would be required to contain dust and noise in a manner that was safe to the Cathedral. The Mora pack was effective in cleaning, but required substantial amounts of water to remove the remains of the cellulose pack. The containment of water and its disposal presented formidable problems within the cathedral, as well as the possibility that excessive use of water might reactivate previous treatments which lie within the dry stone, as has been observed on other projects, and the possibility of staining on the surface due to sustained saturation of the Portland stone.

It was at this stage that discussions with Dr Eddy De Witte of the Institut Royal du Patrimoine Artistique in Brussels led to an evaluation of the recently formulated Arte Mundit® process. This technique is also based on EDTA, but bound into latex. It was noted that the process is quiet to operate, does not create dust, does not require more than very small quantities of water to clean the loosened dirt from the surface, and consequently requires much less protection than other cleaning methods. It also minimizes the possibility of reactivation of past processes should they be lying within the substrate, where excessive moisture absorbed into the surface could mobilize the reactive elements causing variations to the surface and staining. Initial trials on the Portland stone of the interior proved remarkably successful.

Technical development of the project

Following completion of these initial investigations in 1998, the Dean and Chapter, advised by the Fabric Advisory Committee, were satisfied that the project was viable, and authorization was given to develop the project in detail. An initial trial of the selected Arte Mundit® technique was performed over the West Triforium Gallery of the Cathedral. This was selected as the one area where cleaning of a significant area of stonework would not compromise the overall appearance of the interior, should the project for any reason not be able to proceed.

Upon successful completion of this trial area, and following two years of visual monitoring, more extensive cleaning trials and analysis to the stonework were carried out elsewhere in the Cathedral. This was in order to test different conditions, particularly where the stonework was damaged by salts, where it had been structurally repaired in the 1920s with the insertion of new stonework coloured to match adjoining dirty stone, and where heavy greasy deposits stained the stone within the touch of visitors. These trials also extended to vaulted areas and to opposing wall surfaces, including select detailed areas, so that the quality of light within the newly cleaned region could be evaluated (Figures 2 and 3).

Figure 2 A conservator removing the latex film poltice from the stonework in the South Transept. (John Neligan)

Scientific testing and analysis was undertaken by Dr Eddy De Witte to evaluate the technique on the various conditions presented by the stonework in the interior. Detailed analysis was commissioned on the nature of the staining in the Portland stone. Tests were also carried out on the mosaics and on nineteenth-century painted surfaces. Cleaning trials and investigation of the Thornhill paintings in the dome were undertaken by specialist conservators and historic paint specialists. The results of these trials and the research which preceded and accompanied them were discussed at a seminar at St Paul's Cathedral in January 1999. This was attended by representatives of the Cathedrals Fabric Commission for England, the St Paul's Fabric Advisory Committee, English Heritage, The Georgian Group, SPAB, ICOMOS UK, the Corporation of London, the Courtauld Institute and by a number of selected experts in the field.

Following this evaluation, the documentation supporting the project was formalized, and an application for the necessary consent under the Care of Cathedrals Measure was made to the Cathedrals Fabric Commission for England. The project scope of work not only included the cleaning of the interior stonework, but also the painting of the plaster vaults, the cleaning and conservation of existing painted and gilded surfaces, the repair and

Figure 3 A conservator removing the latex film poltice from the stone area of a cherub above the south window of the South Transept. (John Neligan)

cleaning of the mosaics, the cleaning of the monuments (including the substantial Wellington Monument), and – significantly – the conservation of the Thornhill paintings in the dome and the restoration of Thornhill's lost scheme for the tambour. The project also included the relighting of the

Figure 4 The latex film poltice on the walls of the stonework in the north lobby. (Tom Lee, Construction Photography)

interior, an integral part of the whole, as the reflective qualities of the cleaned surfaces significantly changed the quality and indeed the amount of light within the Cathedral. Approval for the project was given in April 1999.

Figure 5 View from the uncleaned Nave, through to completed south lobby and Chapel of St Michael and St George. (Tom Lee, Construction Photography)

Latex films

The only technique useful to clean interiors on large scale and at a reasonable price until recently consisted of EDTA packs. Such poultices allow the removal of thin gypsum layers as well as copper and iron stains. In the last few years several products, all based on the original 'Mora poultice',[1] have been commercialized. Each product consisted of an aqueous solution of EDTA, thickened with a cellulosic derivative. A buffer was also added in order to obtain an alkaline medium. Concentrations of EDTA ranging between 1.65 and 16.7 % (w/v) and pH ranging between 7.8 and 11 have been used. As shown by several investigations calcite substrates also react with the EDTA.[2,3] For a number of calcite substrates, the reaction speed between the poultice and calcite is even faster than with gypsum or metal stains. This can lead to a roughening of surface when the EDTA concentration is too high or the reaction time is too long. Figure 6 shows the influence of the concentration of EDTA on the solubilization of Ca^{2+} on identical substrates. Figure 7 shows the solubilization of Ca^{2+} from different substrates when treated with the same EDTA poultice.

A recent development in interior cleaning is the introduction of peelable poultices for removing superficially adhering dust. This type of product is based on a specially formulated natural latex dispersion. During the evaporation of the water, the polymer is transformed into an elastic film, which adheres slightly to the surface of the stone, attaching to the loosened dust and dirt. The latex can then easily be removed mechanically taking loosened dirt with it, and any residual dirt still on the surface of the stone can be removed immediately with soft sponges and the minimum amount of water.

Physical properties of the latex film poultices

The basic poultice, Arte Mundit® type I, is a specially formulated aqueous dispersion of a natural rubber. The first generation of products contained approximately 0.5 % ammonia, resulting in a pH of 10. A second generation of the product no longer contains ammonia, which makes it possible to apply the products in public areas without evacuating the rooms to be cleaned. In order to allow the removal of chemically adhering pollutants, the original pack can be reformulated with an EDTA compound. The addition of EDTA influences the viscosity of the dispersion, which can be adapted in order to allow the application by spray gun. Depending on the EDTA concentration, the original poultice is available as types II, III, and V. The dry weight including the latex is 60 to 70 % (w/v), depending on the type.

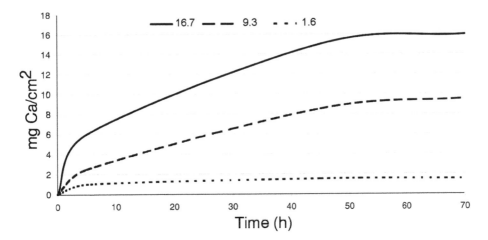

Figure 6 Influence of concentration of EDTA (1.6–9.3–16.7 % (w/v)) and reaction time on the solubilization of Ca^{2+} (substrate – Euville limestone).

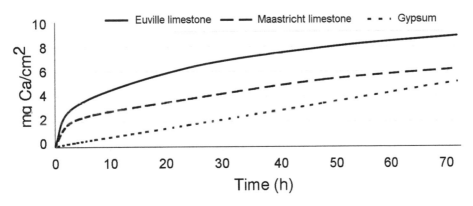

Figure 7 Influence of substrate on the solubilization of Ca^{2+} (concentration of EDTA 10% (w/v)).

Drying rate

The poultice is applied either by brush or specially adapted spray apparatus. The drying process consists of the evaporation of the water and the transformation of the dispersed polymer into a solid but very elastic film. As the product is designed for application in historic buildings, where the interior climate cannot always be controlled, the influence of temperature and relative humidity on the drying time has been examined. The different formulations were applied by brush onto glass sheets, which were then stored under well-controlled climatic conditions. By regular weighing, the evaporation of the water can be followed. Figure 8 illustrates

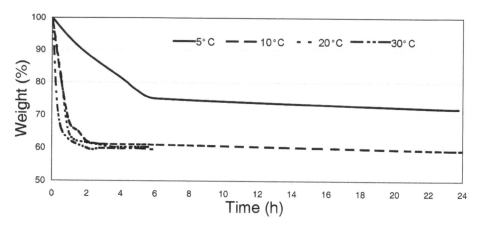

Figure 8 Influence of temperature on the drying rate of the basic poultice at 65 % RH.

the drying rate of the basic product at 65 % relative humidity (RH) and 5 to 30°C.

Between 10 and 30°C, the poultice can be considered as completely dry after 2 to 3 hours. At 5°C the poultice reaches equilibrium after 6 hours. The solid film still contains 10 % of water. Similar tests carried out at 20°C and different RH show that at 95 % RH the drying time increases up to 22 hours before equilibrium is reached. Under normal working conditions, the reaction between the poultice and the substrate stops after a maximum of 2 hours, as there is no chemical reaction between two solids (dry film and substrate). It can even be assumed that the reaction will stop much earlier, as during the drying phase the latex becomes very viscous, which will prevent the migration of the EDTA through the film.

Elasticity

In order to evaluate the elasticity of the dry films, stress-strain tests have been executed. After drying the basic poultice and poultices containing chelating agents, the elongation at break and the strain has been measured at a cross-head speed of 250 mm/min.

Elongation at break of films dried at 20°C and 65 % RH : > 1500 %
Elongation at break of films dried at 20°C and 80 % RH : > 1000 %

At higher RH (up to 95 %) there is a potential risk that the elastic film will not achieve its final cohesion and will tear during removal. This does not seem to have any harmful effect on the substrate to be cleaned, but can make the removal too labour intensive.

Effectiveness

In order to evaluate the effectiveness of the poultices, artificially soiled surfaces have been prepared. Gypsum and calcite slabs have been exposed to either a diesel exhaust, burning candles or treated with a solution of $CuSO_4$ or $FeCl_3$. The slabs covered with the diesel and candle soot were treated with the basic product, whereas the slabs contaminated with the $CuSO_4$ or $FeCl_3$ were treated with a poultice containing a chelating agent. Energy dispersive X-ray (EDX) analysis of a gypsum sample treated with $CuSO_4$ shows that the Cu^{2+} signal at 11 KeV has disappeared after the cleaning process. Figures 9–14 show artificially soiled gypsum surfaces before and after cleaning with the poultice.

SEM-EDX analysis of the inner surface of the poultice, which was in contact with the substrate, showed the presence of the Ca-Na$_2$-EDTA complex crystals (Figure 15). After grinding the film in liquid oxygen, extracting the grinding powder with demineralized water and evaporating the extracting solvent, a Fourrier transform infrared (FTIR) spectrum of the residue is recorded (Figure 16). The obtained spectrum shows all the peaks of the Ca-Na$_2$-EDTA complex.

Figure 9 Gypsum surface covered with diesel pollution (right); a similar surface after cleaning (middle); poultice which was in contact with the pollution (left).

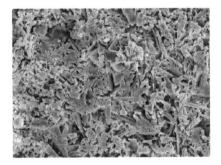

Figure 10 Scanning electron microscopy (SEM) micrograph of a gypsum surface covered with diesel pollution.

Figure 11 The same surface after cleaning; notice the gypsum crystals that became visible after the removal of the soiling.

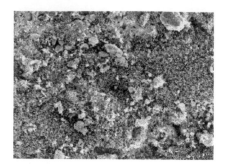

Figure 12 SEM micrograph of a calcite surface covered with candle soot.

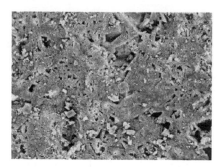

Figure 13 The same surface after cleaning; notice the calcite crystals that became visible after the removal of the soiling.

Figure 14 Euvile limestone contaminated with a copper ($CuSO_4$) stain. Left part cleaned.

Implementation

In order to provide a basis for the proper documentation of the project, a complete digitized photographic record of all internal surfaces of the cathedral was carried out. These photographs were then annotated in order to provide a detailed and quantifiable basis for evaluating the extent of the work; from this bills of quantities were produced, allowing the project to be accurately cost estimated. At this stage, a generous donor offered funding for the entire project, and as a result it was possible to move forward to implementation.

Using the detailed documentation, the project was then put out to competitive tender. A number of possible companies were interviewed in depth as a basis for inclusion on the tender list. Samples of the Arte Mundit® materials were made available to tenderers (none of whom had previous experience of the technique) so that they could evaluate it properly before submitting their tenders. As a result Nimbus Conservation Ltd were awarded the contract for the first phase of the work in April 2001, and work began on site in May 2001.

Figure 15 Ca-Na$_2$-EDTA complex crystals on the inner surface of the poultice after removal from a dried film (black).

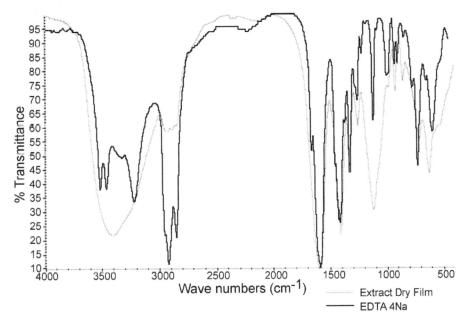

Figure 16 Fourrier transform infrared (FTIR) analysis spectra of the Ca-Na$_2$-EDTA complex (black) and the material extracted.

The work was divided into a series of phases, the contract for each being awarded on the basis of successful completion of the previous section of work, and the contract sum being based on the initial tendered rates (with an uplift for inflation). In the event, Nimbus impressed the design team with their execution and approach to the project, and were reappointed for each successive phase, successfully completing the entire project on time and on budget.

Much of the success in the execution of the work may be attributed to the skill and attitude of the work force. The application of a previously unfamiliar technique on a building of this scale presented a significant challenge, not only of skill and organization, but in retaining the motivation of the conservators involved (and the company is essentially conservator based) over a long period of time in order to achieve a balanced and homogeneous final result. Nimbus carried out the protection of all areas adjacent to those to be treated. They elected to arrange for FTB Restoration (the manufacturers of Arte Mundit®) to carry out the spraying of the latex mix onto the stonework at night – when they could work uninterrupted by the services and events which inevitably punctuated the working day. This also avoided problems of the noise of the spray application. Large surface areas could be sprayed in a two to three night period, leaving the Nimbus team to peel away the latex layer and remove any remaining loosened dirt with small hand held brushes and damp sponges during normal working hours.

Completion

As numerous poultices based on natural rubber latex have been developed, cleaning is either performed through the mechanical removal of adhering contaminants or through chemical removal by chelating with EDTA. Several types have been developed and are available, with variations in the concentrations of the EDTA compound and non-active additives in order to adapt the viscosity to the needs of the spraying apparatus. Under normal conditions the film dries within a relatively short time.

The work on the interior cleaning of St Paul's Cathedral was completed in May 2005. The original computer generated photographic drawings are now being annotated with additional layers of information: these record the extent and nature of the work actually executed, and the archaeological and structural discoveries made during the work. Once completed, these annotated drawings will be lodged in the Cathedral's Fabric Archive.

The project as a whole has brought a radiance to the stonework and indeed to the whole interior of the Cathedral, enabling the architectural elements and the carved detail (of wonderful quality) to be properly seen and appreciated for the first time in perhaps two hundred years. In addition, the project has brought a visual unity to the interior of the Cathedral which it significantly lacked before.

One panel of stonework has been retained uncleaned as a reminder of the Cathedral's previous state. This example was worth retaining as it is now hard to remember the extent of the transformation, so natural does the result appear. Von Uffenbach would probably recognize the quality of light in the Cathedral!

Biography

Martin Stancliffe MA, FSA, Dip Arch (Cantab), RIBA, AABC
Martin Stancliffe founded Martin Stancliffe Architects in 1975. He was appointed Surveyor to the Fabric of St Paul's Cathedral in 1990. In 2004 his company joined the nationally based architectural practice of Purcell Miller Tritton, and Martin became a Senior Principal with the larger practice. He has been a member of the Cathedrals Fabric Commission for England, of the English Heritage Historic Buildings and Areas Advisory Committee, and of the Expert Panel on Historic Buildings and Land for the Heritage Lottery Fund. He was chairman of the Cathedral Architects Association from 1999 to 2005, and is a member of the ICOMOS Cultural Tourism committee.

Inge De Witte
Dr Inge De Witte studied chemistry at the University of Ghent where she finished her PhD in polymer chemistry in 2001. She started her career at FTB Restoration (a company specializing in the production of high quality materials for the preservation and the restoration of wood and stone) in October 2000. Her main tasks consisted of new developments, quality control and all safety issues. She was also responsible for several national and international projects. In June 2005 she stared working at Centexbel (The Belgian Textile Research Centre) as project manager and researcher.

Eddy De Witte
Dr Eddy De Witte is honorary head of department of the Royal Institute for Cultural Heritage in Brussels. He joined the Institute in 1971 where he was head of the Scientific Department from 1987 until his retirement in 2004. His main fields of research related to the long term effectiveness of modern products used in the restoration of historic buildings, and the study of degradation processes of historic building materials. He has been a member of many scientific committees of international conferences, and of national and international commissions for restoration. He carried out missions for UNESCO to Bangladesh and Mostar.

Acknowledgements

The authors wish to acknowledge the assistance of Deborah Carthy of Carthy Conservation; Filip Moens of FTB Restoration; David Odgers and Jenny Jacobs of Nimbus Conservation; Ulrike Knox and Emma Hardisty of Purcell Miller Tritton; and Kyle Normandin, Senior Associate of Wiss, Janney, Elstner Associates New York office, for their technical contributions and assistance with this paper.

Notes

1   Mora, P. and Mora, L., 'A method for the removal of incrustations from limestone and mural paintings', *Problemi di Conservatione* (1973), pp. 339–44.
2   De Witte, E. and Dupas, M., 'Cleaning poultices based on E.D.T.A.', 7th International Congress on Deterioration and Conservation of Stone, Lisbon 15–18 June 1992, pp. 1023–31.
3   Lauffenburger, J., Grissom, C. and Charola, A. E., 'Changes in Gloss of Marble Surfaces as a Result of Methylcellulose Poulticing', *Studies in Conservation 37* (1992), pp. 155–64.

Laser Cleaning of Sculpture, Monuments and Architectural Detail

Martin Cooper

Abstract

The use of laser cleaning has now become routine in a number of specialized conservation studios throughout Europe. The technique is most widely applied to sculpture and monuments, for which commercially available laser cleaning systems are now available and are used to provide sensitive, high-quality cleaning. The use of laser cleaning on buildings has tended to be restricted to areas of sculptural and architectural detail, where cleaning of the highest quality is required. The systems employed on such work are usually 'scaled-up' versions of the systems used in conservation studios, i.e. larger more powerful laser systems which allow faster cleaning rates. Transferring laser cleaning from the conservation studios to the relatively harsh environment of the outdoor work site is not as straightforward as it may seem at first. Considerations such as dirt, power supply, handling of equipment, temperature extremes in summer and winter, and safety must be dealt with if a project is to be completed successfully. Over the past ten years, significant advances have been made in these areas as experience of large-scale outdoor work has been gained and laser cleaning systems have been adapted for such work. This paper describes some of the issues that have been addressed as laser cleaning has moved out of the relative comfort of the conservation studio.

What is laser cleaning?

A laser is a device that produces an intense, pure, and highly directional beam of light (also known as 'laser radiation'). The nature of the laser beam is defined by its wavelength (or colour), its pulse duration (some lasers deliver continuous beams of light), and the energy/power contained within the beam. Lasers used for cleaning sculpture and monuments generally deliver this energy in the form of very short pulses of near infrared light, i.e. energy is delivered to the surface being cleaned at a wavelength of 1064 nm (invisible to the eye) in a pulse lasting approximately 10 ns (or 10^{-8} s). By delivering just a few hundred millijoules of laser radiation to a surface very quickly, it is possible to remove unwanted material in a highly controlled way. The purity of a laser beam allows very selective removal, as some materials absorb the energy very strongly and others only very weakly. The pioneering work of John Asmus and colleagues[1,2,3] in the early 1970s suggested that in fact, under the right conditions, laser cleaning becomes 'self-limiting', i.e. the cleaning action stops 'automatically' once the dirt layer has been removed and the substrate revealed. Consider a hypothetical case with a black pollution encrustation on top of a white marble surface. At a wavelength of 1064 nm (absorption properties are wavelength-dependent), a typical pollution encrustation absorbs energy very strongly and a few pulses may be sufficient to remove a thickness of 0.5 mm. A white marble surface, however, absorbs relatively weakly at the same wavelength and so, provided the conservator knows what he is doing, insufficient energy is absorbed per unit area to damage the marble (most of the energy is harmlessly reflected away). The cleaning process stops even though some pulses may fall on the clean surface, as they are simply too 'weak' to remove any material. The large difference between the absorption properties of the encrustation and the marble means that the laser beam can 'discriminate' to a large extent between the dirty and clean surfaces. It is important to note that it is possible to damage any surface with such a laser beam; although a surface may only absorb a small fraction of energy from the beam, if this energy is focused into a small enough area so that the energy absorbed per unit area is sufficiently high then damage can occur. This point is known as the 'damage threshold' of the material. Cleaning is 'self-limiting' only if the dirt layer can be removed using a laser beam focused to give an energy per unit area below the damage threshold of the substrate. So, as with all cleaning techniques, damage to a surface will result if laser cleaning is carried out poorly.

Research has shown that highly selective laser cleaning can be carried out on a wide range of materials at a wavelength of 1064 nm. Pollution encrustations, corrosion layers, unwanted paint layers, and other coatings have been successfully removed from many materials, including marble, other types of stone, bronze, aluminium, terracotta, plaster, and ivory.[4] Occasionally other wavelengths, e.g. 532 nm (green light), are used to remove materials that do not respond favourably to cleaning at 1064 nm. The Neodymium YAG (Nd:YAG) laser has become the 'workhorse' of laser cleaning systems. This laser has been developed over many years and is now reliable, relatively compact, and robust enough for the conservation studio (Figure 1). The laser beam is usually guided to the surface of the art-work via mirrors within an articulated arm or through a flexible optical fibre (or bundle of fibres). The beam emerges through a hand-held 'pen' and is aimed at the surface by the conservator (Figure 2).[5] Cleaning is as precise as the conservator's hand is steady. The diameter of the laser beam is typically between 1 mm and 20 mm, depending on the energy per pulse delivered by the laser. A focusing lens is often incorporated within the handpiece to produce a diverging laser beam. This means that the beam spreads out as the distance between the handpiece and surface increases. As the beam becomes larger, the energy becomes less concentrated and the cleaning effect therefore becomes weaker. By varying the working distance to the object and the amount of energy in each pulse (through a dial on the laser unit), it is possible to control the beam size and level of cleaning. For a particular laser cleaning system, the cleaning rate can be controlled by adjusting the 'repetition rate', i.e. the number of pulses emitted per second (typically up to a maximum of 30). Cleaning can be carried out very pre-cisely using a small beam and slow repetition rate (e.g. 1 or 2 pulses per second) or relatively quickly using a large beam and high repetition rate (e.g. 30 pulses per second) to 'wash' the light over the surface. It is the role of the conservator to select the parameters appropriate for the object in

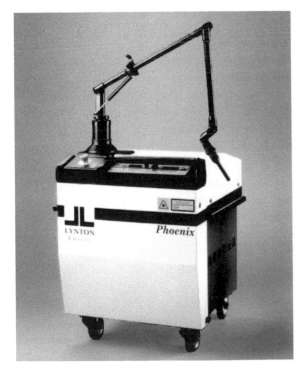

Figure 1 Typical studio-based laser cleaning system (Nd:YAG) with articulated arm delivery system. (Courtesy of Lynton Lasers Ltd)

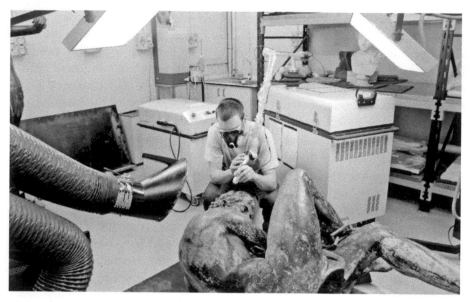

Figure 2 Laser cleaning in a studio environment. Note that the conservator is wearing protective eyewear and a face mask. Lighting is good and a portable extractor unit is used to collect the dirt removed during the cleaning process. (Courtesy of Board of Trustees, National Museums Liverpool)

question, and it is the conservator, therefore, who is responsible for the quality of cleaning that is achieved.

Cleaning with light means that the process is essentially non-contact and well suited to fragile surfaces. The control offered by the technique allows a conservator to remove disfiguring and potentially damaging encrustations, whilst retaining patina and fine surface detail such as tool markings. Laser cleaning is seen as an extremely high quality method of cleaning. Within conservation it is the field of sculpture and monuments that has seen the most widespread use of lasers for cleaning. This includes architectural and sculptural detail on buildings, but not usually complete buildings.

As with all techniques used in conservation, training is important. Appropriate training will allow laser cleaning to be used to its full potential and ensure the safety of both the object being cleaned and the conservator undertaking the work (and other people in the vicinity). Laser cleaning is not a difficult technique to master: conservators can pick up the basics within two to three days. It is then a case of building up experience.

Cleaning in the studio

Laser technology has progressed significantly from the early days when John Asmus put together his first prototype 'statue cleaner'.[6] At that time, lasers were largely unreliable, bulky, and relatively slow at delivering pulses, which meant that although a number of very exciting projects were completed and laser cleaning looked highly promising, the technique was not considered practical or affordable at that time. It was not for another 15 to 20 years that the technology had developed sufficiently for laser cleaning to become a reality in the conservation studio. The last 15 years have seen a slow but steady increase in the use of lasers for cleaning artworks in Europe. There are now about ten manufacturers supplying laser cleaning systems for conservation. In the UK and Ireland there are now seven museums using laser cleaning compared to one ten years ago. All of these are national museums. Probably the greatest hindrance to the uptake of laser cleaning has been the cost of the equipment: a typical studio-based system may cost £30,000 to £40,000. Laser cleaning services are offered by some museums that own their own equipment.

A typical studio-based laser cleaning system is the size of a small refrigerator and weighs about 120 kg. It is easily moved about the studio on wheels and is generally suitable, in terms of size, for cleaning objects that will fit into the studio. The main selling point of laser cleaning is the level of control that it offers the conservator. In terms of cleaning rate, laser cleaning is generally as fast if not faster than other comparable techniques available in the conservator's toolbox. The maximum cleaning rate that can be carried out is determined by the 'average power' of the laser, where average power is the product of pulse energy and repetition rate, and is quoted in watts: e.g. if the maximum pulse energy is 300 mJ (0.3 J) and the maximum repetition rate 10 pulses per second then the average power of the system is 3 W. The average power of a laser cleaning system usually becomes important for large objects (monuments, building façades, etc.) when cleaning rate becomes an important issue. Typical studio-based systems may deliver between 2 and 10 W. Laser cleaning is a 'clean' technique: there is no use of hazardous chemicals; little water is needed (occasionally a small amount of water is brushed/sprayed onto the surface immediately prior to exposure to the laser beam to increase the cleaning rate); and the dust generated during the process is easily collected by an extraction system (Figure 2). In terms of safety, a number of precautions must be taken as the laser radiation being used is potentially hazardous to the eye. The hazards and control measures used to reduce the risk to an acceptable level are laid out in the European safety standard EN 60825-1 (*Safety of laser products. Equipment classification, requirements and user's guide*). It is necessary to put in place safe working procedures and establish a suitable area for laser cleaning work (separated from other areas of the studio, preferably a separate room) to protect other people in the studio.

Cleaning outside

As the use of lasers for cleaning has increased within the conservation studio, so too has their use outside for cleaning monuments and sculptural and architectural detail on buildings. Essentially, the basic process remains the same, with a conservator directing high-power pulses of laser radiation (usually infrared) at the surface being cleaned by means of optical fibres or an articulated arm. The systems employed on such work are usually 'scaled-up' versions of the systems used in a conservation studio, i.e. more powerful laser systems allowing faster cleaning rates. Average power typically varies from 10 to 40 W. Such systems tend to be larger and heavier (typically 200 kg) than the studio-based systems. Lower-power systems can be used on site, but with a corresponding reduction in cleaning rate. At the opposite end of the spectrum an 80 W system has been used in the Netherlands on a number of large-scale outdoor cleaning projects.[7]

Transferring laser cleaning from the conservation studio to the relatively harsh environment of the outdoor work site has added complications in the form of dirt, power supply, handling of equipment, temperature extremes in summer and winter, and safety. Over the past ten years, significant advances have been made in these areas as the experience gained on this type of work has been used to modify laser systems and improve their performance and reliability.

Outdoor work sites tend to be areas with an abundance of dirt and fine dust (often from other work being undertaken), which is not readily compatible with the clean optical components within a laser system. Fine dust has a tendency to fall onto the surfaces of optical components, which can lead to damage requiring replacement of the part unless the system is designed for outdoor work. The main laser unit should be specially sealed to prevent dust ingress as should the articulated arm: mirrors within the arm close to the handpiece are especially vulnerable. It is common practice to cover the joints in an articulated arm with polythene to offer an extra level of protection.

Most laser cleaning systems can be operated from a single phase 110/240-volt power supply, and connection to the mains supply is usually the best option. Unfortunately this is not always possible, and a generator may have to be used as the source of electrical power. Occasionally, the voltage supplied to the laser can be affected by the use of other equipment running on the same supply, which can affect the performance of the laser. It may be necessary for a technician to make some fine adjustments to the electronics inside the laser when the laser is installed on site, based on the number of other pieces of equipment, such as lights or extraction equipment, being used at the same time. Alternatively, it may be preferable for the laser to have a dedicated supply, such as a separate generator.

Laser cleaning systems weighing 200 kg or more have to be moved around on site and special lifting equipment may be required. Systems using optical fibres for beam delivery can remain at ground level and do not require as much handling: a suitable length of fibre can be unravelled and only the handpiece (and control unit) taken to the immediate work area. Generally, optical fibres are more sensitive to damage than the mirrors in an articulated arm and therefore require more maintenance. The laser units themselves are now much more robust than they were ten or so years ago and sockets that once protruded from the back of units (and were susceptible to knocks) have been hidden away and protected. Laser manufacturers have realized that laser cleaning systems will be treated much more harshly on an outdoor site than they will in a contained environment or laboratory.

The possibility of extreme temperatures on site must also be taken into account. If the temperature becomes too hot then the cooler unit may not be able to work efficiently, so it may be necessary to bring additional cooling units on site. This is not usually a big problem, since a conservator would require this anyway to make the working environment bearable. A bigger problem in colder climates, or in winter, is the possibility of the temperature dropping to below freezing. In this situation water inside the laser (used to cool components) can freeze if the laser is turned off. The corresponding volume expansion can lead to damage to optical components and a large repair bill. In recent times, certain manufacturers have incorporated a 'standby' mode that allows cooling water to be circulated (and therefore prevent freezing) even when the laser is switched off. This is particularly useful overnight, allowing the laser to remain on site as long as a power supply is available.

As with working in a studio, laser safety is of paramount importance. On site there may not only be the safety of the conservator undertaking the cleaning to consider, but also other conservators and passers-by. Laser cleaning systems used on site may have a nominal ocular hazard distance (NOHD) of several hundred metres, i.e. the laser beam would have to travel a distance of a few hundred metres before it could be considered too weak to cause damage to the eye. This means that it is necessary to set up a controlled area on site within which the laser beam is confined, or where it can escape without causing injury (Figures 3 and 4). Laser safety on site is therefore about applying the controls and procedures used in a studio to the site so that the risk of injury is reduced to an acceptable level. The procedures may have to be adapted as each situation is usually slightly different and there may be other contractors on site.

The advances in laser technology and experience gained by conservators during the past 15 years have led to significant improvements in the use of lasers outdoors; systems are now more reliable and more robust, and

Figure 3 Laser cleaning site set up on the front façade of St George's Hall, Liverpool. Tin hoarding has been used to enclose the scaffolded area and form a safe working area for laser work. (Courtesy of Board of Trustees, National Museums Liverpool)

Figure 4 Laser cleaning being undertaken on St George's Hall, Liverpool. Note the use of a thick black opaque sheet to divide the laser work area from the non-laser work area. All conservators on site wore safety eyewear while laser cleaning was being carried out. (Courtesy of Board of Trustees, National Museums Liverpool)

higher cleaning rates have been achieved. One of the earliest projects using laser cleaning outdoors on a significant scale was at Amiens Cathedral (France) between 1993 and 1996, where laser cleaning was used in conjunction with micro air abrasive cleaning to remove pollution encrustations from limestone sculpture on the Portail de la Mère Dieu.[8] Since then, laser cleaning has been employed in the conservation of portals of other French cathedrals, including those at Paris, Bordeaux, Poitier, Bourges, and Chartres. This work has led in France to two slightly different approaches to cleaning where lasers are involved: firstly the use of laser cleaning only to optimize the conservation of the stone and its patina; and secondly the use of laser cleaning in conjunction with another technique or techniques in order to obtain an appearance closer to that of the original artwork.[9] Other large-scale outdoor projects where laser cleaning has played a significant role include St Stephen's Cathedral in Vienna,[10] Jeronimos Monastery in Lisbon,[11] the City Hall in Rotterdam,[12] and numerous churches and cathedrals in Italy.[13] Use of laser cleaning on site is still considered by many as a specialized technique, as the cost per square metre is usually higher than for comparable techniques, and its use is therefore often restricted to valuable surfaces where quality of cleaning is of paramount importance. In the year 2000, the entire front façade of Rotterdam's City Hall was cleaned using an 80 W average-power laser. Approximately 2000 m^2 of polluted sandstone was cleaned and cleaning rates approaching 2 m^2/hr were achieved (Figure 5). The system used incorporated a bundle of 45 m long optical fibres, allowing the laser unit to remain at ground level, while cleaning was carried out from a 'cherry picker' personnel lift. This work showed that laser cleaning is feasible on such a grand scale, but projects such as this are still rare.

In the United Kingdom, the idea of using laser cleaning on outdoor sculpture[14] and monuments is still a relatively new concept, with only a handful of practitioners offering the service. The projects completed to date have been very successful. In Liverpool, two high-profile projects in prime locations have recently been completed: the bronze relief panels by Tyson-Smith (approximately 40 m^2) of the Cenotaph on St George's Hall plateau (Figures 6 and 7), and twelve limestone relief panels (approximately 30 m^2 in total) on the front façade of St George's Hall itself (Figures 8 and 9). In both cases, laser cleaning was selected on the basis that it offered the most sensitive cleaning technique. Alternative techniques could have been used to successfully remove the pollution encrustations and unwanted paint and corrosion layers, but these would have resulted in the loss of significant amounts of material from the surface of the artworks themselves, with the loss of surface detail and possibly accelerated weathering.

Figure 5 Façade of Rotterdam City Hall, the Netherlands, during laser cleaning. An 80 W Nd:YAG laser with optical fibre delivery system was used to clean the pollution deposits from the sandstone on the entire front façade. (Courtesy of Martin de Wit, Laserclean b. v., the Netherlands)

Figure 6 The Cenotaph in Liverpool prior to conservation. The bronze relief panels (approximately 40 m^2) were covered by old black paint with areas of active corrosion. (Courtesy of Board of Trustees, National Museums Liverpool)

Figure 7 The Cenotaph in Liverpool after conservation. Laser cleaning was used over a period of about three months to clean the bronze panels. (Courtesy of Board of Trustees, National Museums Liverpool)

Figure 8 One of twelve limestone relief panels (approximately 1.7 m tall) on the front façade of St George's Hall, Liverpool, prior to conservation. (Courtesy of Board of Trustees, National Museums Liverpool)

Figure 9 Limestone relief panels on St George's Hall, Liverpool after conservation. Laser cleaning was used to remove hard black pollution crusts from the weathered stone surface. (Courtesy of Board of Trustees, National Museums liverpool)

What next?

Research and development in the field of laser cleaning of artworks is continuing and leading to the development of novel laser cleaning systems. The LACONA (Lasers in the Conservation of Artworks) conference takes place every two years, bringing together scientists, laser manufacturers, and conservators to discuss the latest research and developments in the field.[15,16] Recent work in Italy has led to the development of a variable pulse length laser cleaning system, which allows the conservator to tailor the duration of the laser pulse to suit the object being treated. Conservators have found that a slightly longer pulse length than the usual 10 ns allows more sensitive cleaning of very delicate surfaces such as the gilded bronze of the Porta del Paradiso by Lorenzo Ghiberti in Florence.[17] Researchers in Crete have developed a laser cleaning system that simultaneously delivers two wavelengths of light (infrared and ultraviolet) to the surface being cleaned. This system has been employed by conservators to successfully remove various forms of pollution encrustation (which cannot be removed satisfactorily at a single wavelength) from marble fragments of the West Frieze at the Parthenon in Athens.[18] However, development of new commercial laser cleaning systems remains comparatively slow. The field of conservation is considered relatively small by manufacturers as compared with the far larger markets of medicine and cosmetic surgery, for example, and therefore tends to rely on developments coming from other disciplines. The relationship between the users and manufacturers is very important as it provides a means of feeding information about the performance of their systems in 'real-life' situations back to the manufacturers. Improvements can then be incorporated into the next system built. Such relationships lead to a constantly evolving product.

As mentioned earlier in this article, laser cleaning is very rarely used to clean the entire surface of a building. It may be chosen as the method of preference for sculptural and architectural detail, but other faster (and therefore cheaper) techniques are likely to be chosen for large areas of flat masonry. The main reason for this is that the laser cleaning systems used for large-scale outdoor work are 'scaled up' versions of the studio systems, requiring a conservator to guide the beam and use their own judgement and experience. This limits the cleaning rate that can be achieved: even with a laser operating at 80 W average power, it might only be possible to reach 2 m²/hr on a polluted sandstone surface. This is significantly slower than alternative 'good quality' cleaning techniques, which might be ten times faster and therefore significantly cheaper. Lasers offering much higher average powers (e.g. 1000 W) can be built, which in theory would allow cleaning rates at least comparable (probably faster) with other techniques. At this point laser cleaning may be seen as an attractive proposition for an entire building. Development is still required to reach this stage. Lasers with an average power of 1000 W emitting 1000 pulses per second would require some type of automated delivery system in order to achieve their full potential in terms of cleaning efficiency (number of pulses required to remove the dirt layer per unit area) and hence maximize cleaning rate. A human being would simply not be able to react quickly enough to scan the laser beam across the surface at the optimum rate. Many pulses would be wasted as clean parts of the surface receive more pulses than necessary. By combining high-power lasers with robotic technology it could be possible to scan the laser beam across the surface at a pre-defined speed that would achieve the appropriate level of cleaning and a significantly increased cleaning rate. Perhaps then the advantages of laser cleaning in terms of sensitivity, selectivity, control, and minimal waste would tip the balance in favour of laser cleaning?

Biography

Martin Cooper
Martin is head of the Laser Technology section at Conservation Technologies. He joined the National Museums Liverpool in 1994, shortly after completing a doctorate investigating laser cleaning of sculpture. He has spent over ten years researching and developing the use of laser cleaning in conservation and is now also actively involved in the three-dimensional recording and non-contact replication of artefacts.

Acknowledgements

The author is grateful to the Office of Science and Technology's PSRE (Public Sector Research Exploitation) fund for financial assistance.

Notes

1   Asmus, J. F., Murphy, C. G., and Munk, W. H., 'Studies on the Interaction of Laser Radiation with Art Artifacts', *Proceedings of The International Society for Optical Engineering (SPIE)*, 41 (1973), pp. 19–27.
2   Asmus, J. F., Seracini, M., and Zetler, M. J., 'Surface Morphology of Laser-Cleaned Stone', *Lithoclastia*, 1 (1976), pp. 23–46.
3   Asmus, J. F., 'Light Cleaning: Laser Technology for Surface Preparation in the Arts', in *Technology and Conservation*, 3 (3) (1978), pp. 14–18.
4   Cooper, M. I., *Laser Cleaning in Conservation: An Introduction*, Butterworth-Heinemann, Oxford (1998).
5   www.liverpoolmuseums.org.uk/conservation/technologies/laserclean2.asp (accessed 26 August 2005).
6   Asmus, J. F., 'The Development of a Laser Statue Cleaner', in *Deterioration of Building Stone, Second International Symposium*, National Technical University, Athens (1976), pp. 137–41.

7   www.laserclean.nl (accessed 26 August 2005).

8   Weeks, C., 'The 'Portail de la Mère Dieu' of Amiens Cathedral: Its Polychromy and Conservation', in *Studies in Conservation*, 43 (1998), pp. 101–8.

9   Bromblet, P., Laboure, M., and Orial, G., 'Diversity of the Cleaning Procedures Including Laser for the Restoration of Carved Portals in France over the Last Ten Years', in *Proceedings of the Fourth International Conference on Lasers in the Conservation of Artworks* (Paris, September 2001), *Journal of Cultural Heritage*, 4 (1), Elsevier, Paris (2003), pp. 17–26.

10 Calcagno, G., Pummer, E., and Koller, M., 'St Stephen's Church in Vienna: Criteria for Nd:YAG Laser Cleaning on an Architectural Scale', in the *Proceedings of the Third International Conference on Lasers in the Conservation of Artworks* (Florence, April 1999), in *Journal of Cultural Heritage*, 1 (1), Elsevier, Paris (2000), pp. 111–117.

11 www.laserclean.nl (accessed 26 August 2005).

12 Ibid.

13 Armani, E., Calcagno, G., Menichelli, C., and Rossetti, M., 'The Church of the Maddalena in Venice: The Use of Laser in the Cleaning of the Façade', in *Proceedings of the Third International Conference on Lasers in the Conservation of Artworks* (Florence, April 1999), *Journal of Cultural Heritage*, 1 (1), Elsevier, Paris, (2000), pp. 99–104.

14 Beadman, K. and Scarrow, J., 'Laser Cleaning Lincoln Cathedral's Romanesque Frieze', in *Journal of Architectural Conservation*, 4 (2) (1998), pp. 39–53.

15 Verges-Belmin, V. (ed.), 'LACONA IV', in *Proceedings of the Fourth International Conference on Lasers in the Conservation of Artworks* (Paris, September 2001), *Journal of Cultural Heritage*, 4 (1), Elsevier, Paris (2003).

16 Dickmann, K., Fotakis, C., and Asmus, J. F. (eds.), 'LACONA V', in *Proceedings of the Fifth International Conference on Lasers in the Conservation of Artworks* (Osnabrueck, September 2003), Springer, Berlin (2005).

17 Siano, S. and Salimbeni, 'The Gate of Paradise: Physical Optimization of the Laser Cleaning Approach', in *Studies in Conservation*, 46, 2001, pp. 269–81.

18 Pouli, P., Frantzikinaki, K. et *al.*, 'Pollution Encrustation Removal by Means of Combined Ultraviolet and Infrared Laser Radiation: The Application of this Innovative Methodology on the Surface of the Parthenon West Frieze', in *Proceedings of the Fifth International Conference on Lasers in the Conservation of Artworks* (Osnabrueck, September 2003), Springer, Berlin, 2005, pp. 333–40.

Conservation of Historic Metals by Waterjetting Techniques

Joseph Sembrat, Patty Miller, Jee Skavdahl and Lydia Frenzel

Abstract

Water cleaning has long been used in the conservation of historic metal artefacts and structures and has been fairly well documented in the conservation literature. Typically, low and medium pressure (not exceeding 4,000 psig) water cleaning techniques such as nebulous mist, steam, and power washing have been used to assist the conservator in implementing conservation treatments, not only because they are safe and effective, but also because conservators are not usually fully versed in the principles and technology behind high and ultra-high pressure waterjetting. High to ultra-high pressure (4,000 psig to 50,000 psig) waterjetting is a specialized technique that has been utilized in the commercial cleaning and surface preparation industry for years but has only been used in the conservation field on a handful of projects. In addition to low and medium pressure cleaning techniques, the authors have used high and ultra-high pressure water cleaning techniques extensively on historic metal artefacts for the removal of coatings, corrosion products and soluble salts while making it possible to preserve desirable original coatings and patina materials.

In metal conservation there exists a delicate balance between the goals of the conservation treatment including surface preparation, with its technical aspects of surface profile or roughness, visible cleanliness, and non-visible cleanliness, and ethical issues of minimal intervention and maximum preservation of original material. Since the late 1980s, conservators have studied the effects of medium and high (not exceeding 35,000 psig) water pressures on bronze monuments in an attempt to replace abrasive blasting techniques to remove unwanted corrosion products while still retaining the aesthetically desirable patina layers. Since then, the authors have performed testing and implemented treatments on steel and aluminium historic artefacts using high and ultra-high waterjetting techniques.

This article will provide a practical introduction to waterjetting technology with specific cases of its use in the conservation of metals. A brief recounting of the application of low to medium pressure waterjetting techniques used by outdoor sculpture conservators will be provided, followed by recent applications of high and ultra-high pressure waterjetting in the conservation of the 'Big Piece', a salvaged steel hull section from the Royal Mail Ship (R.M.S.) Titanic wreck-site and for the conservation of two Saturn V rockets on display at Johnson Space Center in Houston, Texas and the United States Space & Rocket Center in Huntsville, Alabama.

Introduction

As all conservators strive for the best treatment to minimize intervention while maximizing the preservation of original material, there is an ongoing debate regarding surface preparation. For years conservators have used water cleaning to assist in the treatment of historic artefacts. However, these treatments utilized low to medium water pressures and relegated larger scale surface preparation projects to dry abrasive media blasting. In the metals conservation field, high and ultra-high pressure waterjets are rarely used and there are very few testing or application cases documented. In 1992, following detailed engineering and structural studies performed by the National Aeronautics and Space Administration (NASA) and the commercial aircraft industry, high and ultra-high pressure waterjets were adopted for use in metal surface preparation for its ability to remove coatings while retaining the original profile on aluminium aircraft bodies.[1] Its effectiveness was not limited to coated surfaces but was found to also be excellent in cleaning critical engine parts.[2]

For conservators employed in the treatment of historic metal artefacts, a natural progression of this logic indicated that high and ultra-high pressure waterjets would also be appropriate for large-scale industrial artefacts such as ships, aircraft and machinery. Through the use of waterjets, conservators proposed that high and ultra-high pressure water would clean more efficiently and potentially be capable of retaining original material and desirable patina far better than abrasive blasting.

Given the scarcity of existing documentation on the use of higher pressure waterjet cleaning in conservation, the authors brought together prominent members of the waterjetting and conservation communities to sufficiently test the effectiveness of waterjetting for surface preparation on historic metals, more specifically steel and aluminium. High and ultra-high pressure waterjets were chosen for both the conservation of the 'Big Piece', a salvaged 17-ton steel hull section from the Royal Mail Steamship (R.M.S.) Titanic wreck-site and for the conservation of two

Saturn V rockets on display at Johnson Space Center (JSC) in Houston, Texas and the United States Space & Rocket Center (USSRC) in Huntsville, Alabama. The authors were successfully able to remove the mostly non-original coatings while preserving some original coatings (including primers), corrosion products and mitigate both visible and non-visible soluble salts. By adjusting the waterjetting parameters the original coatings, patinas, and surface profiles[3] were left undisturbed. In many places, ultra-high pressure waterjets prepared the surface for coating without the addition of abrasives or chemicals.

Waterjetting technology

Waterjetting, a process harnessing the power of water when its velocity is increased, first appeared within the engineering community in the 1960s. Dr Norman Franz, a forestry engineer and professor at the University of British Columbia, was looking for a faster method of slicing large trees into lumber. Franz discovered he could cut wood if he dropped large weights into a column of water, forcing the pressurized water through a small orifice at the bottom of the column.[4] The advancement of the technology in the second half of the century led to the development of equipment that can not only create these ultra-high pressures (UHP) but sustain them in a continuous stream for extended periods of time. UHP waterjetting technology is comprised of highly specialized equipment that has been in a constant state of development and improvement for the past 20 years. At a minimum, a UHP waterjetting system must have the following components: UHP pump; UHP hoses; tumble box and jetting tool; nozzle; and control hoses. These are the mechanical tools that create the energy needed to push water through an orifice at 35,000 plus psig.[5] To date, there are many industrial contractors that specialize in the development and application of UHP waterjetting.

Figure 1 UHP waterjetting equipment.

Waterjetting systems are not glorified power washers; they are refined systems that require a balance between several parameters, all of which are adjusted depending on the substrate, coatings, and corrosion products being treated. Pressurized water has two working components:

- direct impact or erosion effect that is principally controlled by the pressure of the stream
- hydraulic or lifting effect that is principally controlled by the volume of the stream

When undertaking surface preparation operations which require the use of UHP, a firm understanding of the principles of the equipment and how it performs including a systematic testing program is required to properly determine the appropriate balance of the following variables:

- pressure – as measured at the gauge (psig)
- nozzle design
- orifice size – diameter of the opening in the nozzle for the water stream
- angle of incidence – the angle at which the water stream impacts the surface
- stand-off distance – distance of the nozzle to the working surface
- dwell time – the length of time that the work surface remains in the path of the water stream

Pressure and orifice size determine the velocity of the water impacting the surface. Pressure and velocity are directly related (i.e. increased pressure, increased velocity), whereas orifice size and velocity are inversely related (i.e. decreased orifice size, increased velocity). Increasing the orifice size will increase the hydraulic force of the water exiting the nozzle and has a greater potential to deform thinner, more delicate metals. Definitions of pressure vary depending on the industry, but conservation professionals have adopted the following criteria for applications to metal substrates which are based on the conservation expertise and surface preparation/waterjetting industry standards including The Society for Protective Coatings (SSPC), National Association for Corrosion Engineers (NACE) and WaterJet Technology Association (WJTA):

- low pressure (LP) < 800 psig
- medium pressure (MP) = 1,000–4,000 psig
- high pressure (HP) = 4,000–30,000 psig
- ultra-high pressure (UHP) > 30,000 psig

Nozzle designs will affect the quality and speed of the treatment. It is imperative that an appropriate nozzle is chosen for each waterjetting task. Fan-jets have a concentrated stream whereas a rotating nozzle utilizes a concentrated stream that is broken into multiple streams. Waterjetting above 30,000 psig requires the use of a rotating nozzle, as a stable fan-jet has not been designed to withstand ultra-high pressures.

Waterjetting streams impact the surface with an angle of incidence. A direct stream perpendicular to the surface has an angle of incidence of 0 degrees. Directing the jetting tool at an angle from the perpendicular will decrease the amount of force being applied to the part of the stream furthest from the surface and increase it in the area closest to the surface. Although not standard practice in the waterjetting industry, in some cases this effect might be useful as it would greatly reduce the amount of erosion while removing loosely adherent materials.

Each waterjetting nozzle performs differently and has its own optimum stand-off distance. This is a variable controlled by the operator or 'jetter' which provides him a method of increasing or decreasing the force of the waterjet. A more sensitive area of the work surface could be treated accordingly using the same parameters simply by increasing the stand-off distance without changing the other variables. This may offer an advantage when working on substrates with varying degrees of stability or one that is composed of different materials.

The amount of material removed is directly related to the dwell time. This provides the flexibility of selectively removing material and/or separating one layer from another (i.e. topcoat and primer).

Although all treatments have their limitations, the parameters involved in waterjetting provide a flexibility that is advantageous in a field that does not often see uniform surface conditions or objects that are composed of a single material. The parameters can be adjusted as needed to successfully treat the surface. A painted metal surface treated with waterjets will be left free of soluble salts, corrosion products, and coatings with no deformation of the existing surface profile. Surfaces are visibly and microscopically clean and, since this is a 'cold' cleaning process, a waterjetted surface will have no heat affected zone.

There is a significant lack of published conservation literature about the use of medium to ultra-high pressurized water blasting. Two papers of note date from the late 1980s and early 1990s. These papers, written by materials specialists and conservators, present the advantages of using waterjetting techniques for the conservation of outdoor bronze sculptures. With the focus being to remove corrosion from pits in the bronze, both papers discuss the superior cleaning results obtained using pressurized water compared to abrasive techniques.

Merk-Gould's paper[6] is written from the conservator's view and is a comparative study of abrasive cleaning techniques in use in the late 1980s to early 1990s and of the potential to use medium pressure water (2,000–4,000 psig) for the removal of corrosion from outdoor bronze sculpture. The authors show that medium pressure water produces a controlled erosion of the corrosion layers from the bronze surface. Detailed in the paper is a waterjetting process referred to as micro-tunnelling, the flooding of microscopic surface pits using medium pressure water. The paper then goes on to document that this micro-tunnelling removed more of the green sulphate corrosion than abrasive processes using walnut shells and bronze wool while preserving the desired metal patina.

Draughon's paper[7] is a general introduction to ultra-high pressure waterjet cleaning and describes the field case study of cleaning the William Penn Statue on top of the Philadelphia City Hall with ultra-high pressure waterjets in the late 1980s. It is important to remember that the mechanisms of cleaning by water or by solids are two different processes. Abrasives will always change the surface characteristics of the surface against which they are propelled; as the solids hit, abrade, and/or erode the substrate, the surface profile is altered. This is the nature of how abrasives work in the 'cleaning process'. When abrasives hit the surface the ductile metal flows and/or is moved around, thus changing the substrate profile. Dirt or unwanted material in crevices and cracks cannot be reached by a particle that is larger than the crevice opening, whereas water is not limited by the size of the pits or crevices. The mechanism of micro-tunnelling will increase the cleaning effect into areas smaller than the abrasive particles. These phenomena are described in detail in Dr Summers' book *Waterjetting Technology*, a canonical reference in the waterjetting industry.[8]

Although secondary to the preservation of the artefact, decontamination of a site/equipment is a large task that often requires a significant amount of labour and can drain a project of its resources. Aside from the task of capturing wastewater under and around an artefact, waterjetting presents several clear advantages over alternative surface preparation procedures.

Figure 2 Photomicrograph of a clean and dirty surface of Alexander Milne Calder's William Penn statue on top of Philadelphia's City Hall taken before (left) and after cleaning (right). (Ed Oechsler)

The choice to use waterjetting will greatly reduce the exposure of workers to known hazards associated with abrasive blasting (particulate matter) and procedures requiring chemical solvents (respiratory/skin irritants, carcinogens). Waterjetting does not independently contribute to the hazard level of the wastewater generated. This is determined by the coatings and/or corrosion products removed from the substrate, as well as any chemical additives used in assisting the jetting process. Coating removals (i.e. lead paint) performed in conjunction with chemical additives may require filtration, after which the wastewater is safe and can be dispersed directly into the sanitary sewers.

The following two case studies describe how ultra-high pressure waterjets were selected and used in the conservation of two historically significant artefacts.

Case study: R.M.S. *Titanic*, 'Big Piece'

History

On 31 May 1911, the black hull of the *Titanic* was launched after 15,000 shipbuilders laboured two years to build the largest moving object made by man. When fitted, the liner was 11 stories high and almost three football fields in length.

On Sunday 14 April at least six warnings from neighbouring ships would alert the *Titanic* of icebergs and field ice in her vicinity. At 11.40 pm she struck an iceberg. By 2.20 am on Monday 15 April, the *Titanic* sank in the cold Atlantic Ocean. By 4.00 am the nearby ship *Carpathia* reached the wreck-site. Only 705 passengers out of 2228 would survive.

The *Titanic* presently rests 3,798 metres (12,460 feet or 2.5 miles) below the surface of the North Atlantic Ocean, 725 kilometres (450 miles) southeast of Newfoundland. The two main parts of her hull are separated by 610 metres (2,000 feet). Apparently, the *Titanic*'s hull had broken on or near the surface. The huge engines plummeted quickly to the ocean floor, while the bow and stern sections spiralled to points some distance away. As they spiralled downwards, their contents scattered, forming what is referred to as the debris field.

The 'Big Piece' which was recovered on 10 August 1998 and aptly named for its considerable size and weight when compared to other recovered artefacts, was approximately 16 kilometres (10 miles) from the rest of the *Titanic*'s wreck site, where it had drifted after an unsuccessful recovery attempt in 1996. The piece is believed to be from two empty First Class suites, C-79 and C-81, which were located on the starboard side of the ship. These cabins were next to the staterooms of New York theatrical producer Henry B. Harris and W.T. Stead, the most famous journalist in

Figure 3 The 'Big Piece' as it appeared prior to its recovery from the R.M.S. *Titanic* wreck site. (R.M.S. *Titanic*, Inc.)

England at the time. Both Mr Harris and Mr Stead lost their lives aboard *Titanic*.

Before the 'Big Piece' was separated into two smaller sections to allow for easier handling and transport, it measured approximately 8 metres wide × 6 metres deep (26 feet × 20 feet) and weighed approximately 20 tons. It consists of three steel plates, each approximately 2.5 cm (1 inch) thick, which are riveted together and at one time formed a single watertight skin. Four bronze portholes, three of which retain glass and a lead methane pipe are part of the piece. In addition, remnants of original paint still survive in several locations.

While the hull was submerged in saltwater in the North Atlantic Ocean, it absorbed salts (sodium chloride) and became weakened, stained, and encrusted with rusticles, a complex mineral compound excreted by a community of micro-organisms which feed on the rusting iron. Upon bringing it to the surface in 1998, the metal had to be kept wet to reduce the rate of corrosion until the desalination process could begin; however, it was also to be immediately exhibited. Because of the requirement that the piece had to be displayed while the desalination process was underway, it was initially exhibited in Boston under a less than ideal shower spray system consisting of a solution of sodium carbonate and water. This system was ineffective in removing salts from the piece and was discontinued immediately after the Boston show.

Once the 'Big Piece' left Boston, it was moved to two other exhibition sites where it was submerged in a large above ground swimming pool filled with a solution of sodium carbonate and water that provided a suitable soaking vessel. A sacrificial anodic system, designed by Mesa Products, was attached to the surface of the piece. The combination of the electrolytic solution and the anodic system provided an electrical path allowing the sodium and chloride ions to move out of the steel, and attack the aluminium and magnesium anode blocks fastened to the hull.

After soaking in the solution for a period of 20 months, the piece had to be prepared for an upcoming exhibit. In December 1999 the massive hull section was suspended from a mobile gantry system so that all sides of the piece were made easily accessible to the conservators. Once properly suspended, the hull was water blasted with 3,000 psig water to remove loose corrosion and unstable rusticle incrustations but retain as much original patina as possible. The remaining corrosion products and incrustations were stabilized with a solution of tannic acid and water and the entire piece was then coated with a commercially available protective barrier coating of microcrystalline wax (Briwax) to help reduce formation of corrosion in future.

The 'Big Piece' then began to travel on exhibit for the next four years. It was transported, displayed, and stored in non climate-controlled environments. Fluctuations in temperature and humidity caused the once firmly attached rusticle incrustations to delaminate from the surface. During periods of travel the deposits absorbed moisture and swelled. Once the piece was moved into an environment with low humidity, generally during periods of display and storage, the deposits dried out and shrank. This phenomenon of repeated swelling and shrinking gradually resulted in the detachment of deposits from the surface. Touch-up treatments were performed between exhibition periods; however, by the summer of 2004, the 'Big Piece' was in need of a complete retreatment.

The six major objectives of the 2004 conservation treatment of the R.M.S. *Titanic* 'Big Piece' involved:

- removing the existing deteriorated protective wax coating
- removing the delamination layer (rusticle incrustations)
- removing as much soluble salts as possible
- stabilizing surviving original paint
- converting the existing corrosion and seal the surface
- stabilizing loose rivets and subcomponents for shipping and display purposes

Removal of the unstable rusticle incrustations and loose corrosion products

The goal of the conservation treatment was to remove the unstable rusticle incrustations and loose corrosion layers without exposing bare metal, eroding the metal surface, or destroying the desirable patina of the piece.

Figure 4 Detail image showing the delamination of the rusticle incrustation.

Based on the success of the initial treatment of the piece in 1999 and other significant treatments which involved the use of waterjets, it was decided that this approach would be tested to determine its viability for this project. Several tests were performed to determine the effectiveness of higher pressures on the removal of the rusticle incrustations. Because the initial treatment had been completed with pressures not exceeding 3,000 psig, this was the first pressure tested. With heat (180°F/83°C), 3,000 psig was effective in removing the protective wax coating; however, it did not remove the remaining rusticle incrustations in a consistent manner.

A team of waterjetting professionals and conservators convened to determine the range of pressures that could be safely used to achieve the intended goals. Testing proved that 10,000 psig delivered through a rotating nozzle with an orifice size of 0.002 inch used at an angle of incidence of 0°, 2 to 3 inches from the surface,[9] and at a dwell time of approximately 1 sec/in^2 was capable of removing soluble salts and the majority of the incrustations on the flat surfaces and that 20,000 psig, using the same operating parameters as with the 10,000 psig, was needed to remove the thicker crusts which were impacted around the edges of the rivet heads. The combination of these two working parameters successfully avoided damage to the metal substrate. This also greatly reduced the amount of time conservators would have to spend removing the remaining crusts by hand using dental picks.

Figure 5 Surface after waterjetting with both 10,000 and 20,000 psig. (Steve Dunne)

Figure 6 High pressure waterjetting. Inset picture showing the rotating nozzle and its associated jewels. (Steve Dunne)

Soluble salt removal

In order to maximize the amount of salts that would be cleaned from the piece, water was flushed between plates, causing the salts within to become mobilized. Runoff water was captured at specific locations and throughout the cleaning process, then chloride readings were taken using Merck Quantitative test strips and the more sensitive Elcometer 134W test. This process was repeated throughout the cleaning process and the results were graphically recorded. Ultimately, it was found that the chloride levels were reduced from over 1,000 mg/l to below 25 mg/l.

Conversion of the stable corrosion products

The stable corrosion products that remained on the surface of the piece were treated with Fertan,[10] a proprietary blend of tannic acid and phosphoric acid. Once the Fertan had fully reacted, it was rinsed from the surface and the piece was allowed to dry.

Dewatering of the piece and sealing the surface

When water is driven into crevices under pressure, it must be dried out or evaporated before a coating or sealant is applied. Typically, an industrial waterjetting contractor would use pressurized air to drive out trapped water from areas where support brackets meet or other areas where water can be trapped. In the case of the *Titanic*, water was found to be weeping out of pores in the metal well after the surface moisture had evaporated. Using pressurized air to force water out of all the pits and holes in the metal was an unrealistic approach, so the conservators used a combination of heat to drive moisture out and away from the metal with a climate-controlled plastic enclosure around the piece to dehumidify the space to a range of 40% to 50%.

Once the piece was completely dry and there was no evidence of any weeping, the surface was heated to approximately 200°F/94°C and received a protective barrier coating of a wax mixture containing a multi-metal corrosion inhibitor. The wax mixture was based on the United States' National Park Service formula and is composed of (percent by weight): 71% Victory Brown, 13% Polywax 1000, 13% Petranauba C, and 3% Polywax 500. The liquid corrosion inhibitor, Cortec M-238 Multi-metal Corrosion Inhibitor,[11] was added at 5% by volume to the final mixture.

Figure 7 Detail image of a rivet head showing water weeping from pores in the metal. (Steve Dunne)

Figure 8 Overall image of the 'Big Piece' after the conservation treatment. (Steve Dunne)

Case study: Saturn V rocket

The conservation of two Saturn V rockets is occurring concurrently: one is located at the Johnson Space Center (JSC) in Houston, Texas and the other at the United States Space & Rocket Center (USSRC) in Huntsville, Alabama (adjacent to NASA's Marshall Space Flight Center). These are two of the three surviving vehicles built to launch Americans to the moon during the late 1960s. Measuring 110.64 metres (363 feet) long (the Statue of Liberty is approximately 18.92 metres (60 feet) shorter, even when including its base) and capable of generating 3.4 million kg (7.5 million lbs.) of thrust, the Saturn V remains the largest, most powerful American launch vehicle ever built.

The Saturn V was manufactured using various types of metals including aluminium alloys (extruded, cast and milled), aluminium honeycomb sandwich, stainless steel, and titanium alloy. Non-metals that make up exterior surfaces include spray-on polyurethane foam, cork, and various types of plastic, phenolic resin, and fibreglass composite. Aluminium (alloys 2014, 2219 and 7075) is the most prevalent of the metals used to construct the large diameter fuel tanks, which comprise the majority of the vehicle, and the extremely thin skirts, which provide structural support and transition points between the rocket stages.

Saturn V at Johnson Space Center

Displayed near the entrance of NASA's Johnson Space Center since 1977, the JSC Saturn V has suffered from long-term exposure to the Houston climate of high humidity, high temperatures, high ozone concentrations, salt air, and chemical pollution which has caused extensive corrosion of metals, especially the aluminium alloys, and degradation of all other susceptible materials. Previous attempts to 'spruce up' the rocket introduced additional problematic conditions, which resulted in further deterioration and

Figure 9 Saturn V Rocket located at Johnson Space Center before treatment.

Figure 10 Overall image and insets of Stage II before treatment, showing deterioration of polyurethane foam and aluminium alloys.

loss of original materials. Rapid deterioration of the rocket was evident within five years of being placed on display at JSC. Peeling and blistering paint and corroding metals prompted NASA to intervene, allowing extensive sand-blasting of the rocket surfaces and removing paint down to the bare metal. Not only did this remove the original protective chromate treatment on the aluminium surface it also spread soluble salts over previously uncontaminated areas of the vehicle. Today the metal is heavily pitted and exfoliating, covered with up to ten layers of paint, and is suffering from a broad range of corrosion.

Saturn V at US Space & Rocket Center

In contrast, the Saturn V located at the US Space & Rocket Center, whilst on outdoor display for nearly seven years longer than the JSC rocket, is far better preserved. Its current state of preservation is due to a less aggressive climate, a more stout construction (as a testing vehicle as opposed to a flight vehicle), and less invasive maintenance. Paint coating testing identified the presence of original zinc chromate primers and original paint layers throughout. Subsequent paint coatings numbered from four to as many as fourteen layers. Corrosion was almost completely limited to areas where animal debris and faeces had collected over the years and areas where water pooled on the surface.

Project scope and the goals of the treatment

The general scope of the two projects consisted of the following tasks:

- conduct a comprehensive condition assessment of each vehicle to fully document present conditions and perform historic research
- implement a testing and analysis program to determine the condition of the materials, identify root causes of the observed deterioration, and provide appropriate methods of intervention and repair to reduce or arrest the rate at which the materials are deteriorating
- general cleaning and removal of bulk debris and micro-organisms
- repair and stabilization of damaged and corroded metal components including surface preparation, metal repairs, and coating application
- repair and stabilization of non-metal components including the treatment of phenolic resins, urethane foam, cork, electrical wiring, mylar, rubber, and the abatement of all hazardous materials

The intended treatment goals were ones that emphasized minimal intervention by using the least aggressive means possible to achieve the most successful conservation results. It was an approach that required trying to balance performance and longevity of the treatment with the ethics and standards of the conservation field of practice – a very daunting task considering the enormous size and complexity of the two objects. In order to undertake two projects of this magnitude simultaneously and within a budget established by the clients, it was necessary for the conservators to identify the major issues affecting the vehicles and a means to effectively conserve them. Early on it was determined that the greatest challenge would be how to effectively remove coatings from thin aluminium skin while retaining the original profile (and as much original primer as possible on the USSRC vehicle), and avoid causing distortion to the underlying metal substrate. The primary goal on the JSC rocket was to retain as much of the original coating system as possible. However, surface pull-off adhesion tests (ASTM D 4541 – Pull-off Elcometer) revealed that extensive underfilm corrosion in numerous locations was leading to unacceptable pull-test readings (<50 psi). Additionally, archival research and paint analysis confirmed that the majority of the original coatings had been abrasively blasted from the rocket during an overly aggressive 1983 restoration project. Unfortunately, abrasive blasting did not remove soluble salts, but instead created more crevices and pits in the metal surface for salts to be trapped.

There is much that could be discussed regarding to the overall treatment of these two vehicles. However, the focus of discussion here relates to the decision to use ultra-high pressure (UHP) for the removal of coatings on the vehicle, and the methods and materials actually used to implement the treatment.

Testing of ultra-high pressure waterjets

Based on our extensive use of water-based cleaning techniques for surface preparation of metal substrates on past projects, the comprehensive coating removal research that had been performed by the aircraft industry, the stringent environmental restrictions placed on the project team by the two sites where the work was to be performed, and the need for an effective means of removing multiple layers of paint and soluble salts, led us to evaluate the potential use of UHP for these two projects.

Conservators collaborated with material specialists and the client to develop a comprehensive testing programme aimed at establishing the appropriateness of the technique and the proper operating parameters for cleaning, coating removal, and salt passivation/extraction for the surface preparation of the metal surfaces prior to repair and repainting. In the spring of 2004, sections of the historic aluminium skin and honeycomb panels from the JSC Saturn V were exposed to waterjetting pressures ranging from 3,000 psig to 50,000 psig. They were also subjected to a variety of stand-off distances and tip orifice sizes in an attempt to determine the various operating parameters that would be required to achieve different levels of surface preparation. In addition, heated water, various rinse aids, chemical paint strippers, and in-stream abrasive materials were tested in conjunction with the waterjetting. Aluminium surfaces before and after cleaning were tested using the Chlor*Rid Test Kit Chlor*Test 'CSN Salts', a surface preparation industry standardized test kit used to quantify chloride/sulphate/nitrate ions present on surfaces prior to repainting of metals materials. Samples were removed from coupons and sent to a laboratory for testing (SEM/EDS) to identify and quantify residual surface contaminants (Table 1).[12]

The results of the testing for the JSC rocket can be summarized as follows:

1 Where existing coatings were to be substantially removed to the metal substrate, 20,000 psig to 40,000 psig water delivered through a rotating nozzle equipped with four – 0.001 inch jewels and used at an angle of incidence of 0°, 2 to 3 inches from the surface, at a dwell time of approximately 1 sec/in^2 was ultimately recommended.[13]
2 Pressurized water alone was sufficient to remove non-visible contaminants without addition of cleaning agents in the water stream (e.g. not detected by field methods or below NV-2 of SSPC SP 12).[14]

Sample Number	Material	Treatment Process	Test 1 (5000X)	Test 2 (5000X)	Test 3 (5000X)	Comments
64-501	Aluminium	control	C, O, Na, Al, Cl, Ca, Zn	C, O, Na, Al, Si, Cl, Ca, Zn, –P	C, O, Na, Al, Cl, Ca, Fe, Zn	Chlorides and silicates present
64-502	Aluminium	control	C, O, Na, Mg, Al, Si, Cl, K, Ca, Ti, Cr	C, O, Na, Al, Cl, Ca Zn, –Cu	C, O, Na, Al, Cl, K Ca, Zn, –Cu	Chlorides and silicates present
64-503	Aluminium	20K wj				
64-504	Aluminium	20K wj				
64-505	Aluminium	20K wj/1:5 Cortec VpCI-427 (3 min exp)	C, O, Al, Cu, Zn –Mg	C, O, Al, Cu, Zn, –Mg, Re	C, O, Al, Zn, –Mg	Re present and not in controls
64-506	Aluminium	20K wj/1:8 Cortec VpCI-427 (3 min exp)	C, O, Al, Si, Cu, Zn –Mg	C, O, Al, Fe, Cu, Zn, –Mg, S	C, O, Al, Si, Cu, Zn, –Mg	Sulphates and silicates still present
64-507	Aluminium	20K wj/1:15 Cortec VpCI-427 (3 min exp)	C, O, Al, Cu, Zn –Mg	C, O, Al, Zn, Mg –Cu	C, O, Al, Zn, Mg –Cu, Cr	chromates present
64-508	Aluminium	20K wj/Eldorado AC-12 1:1 (5 min exp)				
64-509	Aluminium	20K wj/Eldorado AC-12 1:1 (10 min exp)	C, O, Al, Cu, Zn –Mg	C, O, Al, Cu, Zn	C, O, Al, Cu, Zn, –Mg	
64-510	Aluminium	20K wj/Eldorado AC-12 1:1 (15 min exp)	C, O, Al, Si, Zn, –Cu	C, O, Al, Cl, Fe, Zn, –Mg, P	C, O, Al, Cr, Zn	Chlorides (minimal) and silicates present
64-511	Aluminium	20K wj/Undiluted Sea 2 Sky SPC-501 (2 hr dwell)	C, O, Al, Cu, Zn –Ge	C, O, Al, Cu, Zn, –Ge	C, O, Al, Zn, Ge	Ge present and not in controls
64-512	Aluminium	20K wj/Undiluted Sea 2 Sky SPC-501 (3 hr dwell)				
64-513	Aluminium	20K wj/Undiluted Sea 2 Sky SPC-501 (3 hr dwell)	C, O, Cr, Zn, Br –Mo, Fe	C, O, Cr, Zn, Br –Mo, Fe	C, O, Al, Cr, Zn –Mo, Fe	chromates present; minimal sulphates present
64-514	Aluminium	20K wj/1:15 Cortec VpCI-427 (3 min exp)	C, O, Al, Zn –Mg, P	C, O, Al, –Zn, Mg, Cr	C, O, Al, P, Zn –Mg	
64-515	Aluminium	20K wj	C, O, Al, Si, Cl, Ca, Cr, Cu, Zn –S	C, O, Na, Al, Si, Cl, Ca, Cr, Zn –S, Re	C, O, Al, Si, Cl, Cu, Zn –S	Chlorides, sulphates and silicates still present; Re present and not in controls
64-516	Aluminium	20K wj	C, O, Al, Si, Cl, Ca, Cr, Cu, Zn –Mg, S	C, O, Al, Si, Zn, –Mg	C, O, Al, Si, Cl, Cr, Fe, Cu, Zn, Ge	Chlorides, sulphates and silicates still present; Ge present and not in controls

Table 1 Metals sample testing on aluminium components removed from Stage III of the Saturn V Rocket located at the Johnson Space Center (see also note 12).

3 Pressurized water can be used to remove the coatings from a minimal extent to the bare substrate.

4 A clean shear between coatings did not occur.

5 Heating the water appeared to increase the effectiveness of the removal of coatings.

6 Heating the water appeared to increase the effectiveness of the removal of salts as determined by the comparison of conductivity between ambient and hot water cleaning at 3,000 psi.

7 Chemical softening strippers followed by pressured water were very effective in removing the coatings. In practice, it could be difficult to prevent removal at an intermediate layer.

8 The water-abrasive combination was very effective in coating removal to the bare substrate but was found to be too aggressive to prevent removal of intermediate layers.[15]

9 In areas where extensive corrosion existed and where metal repairs were to take place, low pressure water cleaning in combination with phosphoric acid-based metal brighteners would be used.

Following the same guidelines established for the JSC vehicle, samples of the aluminium skin from the USSRC Saturn V were similarly tested in the spring of 2005. Testing determined that it was possible to safely remove multiple paint layers from the rocket skin while retaining the original

Figure 11 Detailed image of epoxy manufacturer stamp on the USSRC S-IVB tunnel cover.

tightly adhered zinc chromate primer using up to 25,000 psig and the same parameters as described for the JSC rocket. Original epoxy manufacturing stamps applied over original primers as well as applied directly to the metal surface were uncovered during testing and not affected by the force of the jets. This was considered to be a great success as it allowed the authors to achieve their surface preparation goal of removing all finish coatings and preserve as much of the tightly adhered original zinc chromate primer as possible. Soluble salt levels overall were in the very low end of the field detection range (chlorides and nitrates detected below 30 ppm, sulphates at below 5 ppm) as indicated by the Chlor*Rid Test Kit Chlor*Test 'CSN Salts'. Waterjetting further reduced salt levels to a non-detectable range of below 5 ppm for chlorides, and below 0 ppm for nitrates and sulphates. Severe corrosion would be treated by mechanical removal and the use of chemical brighteners, much the same as they were to be treated on the JSC Saturn V.

Implementation at the project level

Ultra-high pressure waterjetting was selected as the primary surface preparation method for both Saturn V projects because it offered a great deal of control in removing unstable coatings to a stable painted surface or to the bare metal, without removing or damaging the original surface of the aluminium substrate. Where possible, conservators specified that coatings be removed to stable paint coatings. For the JSC Saturn V, complete coating removal was ultimately specified as there was little or no original paint present, there was extensive underfilm corrosion throughout the vehicle, and it would provide the most effective means of eliminating corrosion-causing soluble salts. In the case of the USSRC Saturn V conservators specified that coatings be removed down to the original zinc chromate primer layer. Accordingly, this approach of selective coating removal would not be practical in all locations as the thickness and stability of the primer was not consistent on all surfaces. Therefore it was necessary to remove coatings down to the bare aluminium surface in some locations.

In July 2005, coating removal and cleaning using waterjets began on both the JSC and USSRC Saturn V rockets. On-site implementation of the specified waterjetting techniques revealed that initial tests of pressures of 25,000 psig were sufficient to remove multiple paint coatings down to the zinc chromate primer layer. However, 35,000 psig was found to produce the most efficient cleaning rate on both projects for tank and skin surfaces, requiring the operator to dwell less time in sensitive areas and therefore, leading to less opportunity for operator induced damages. However, on the USSRC Saturn V, this pressure removed more of original primer than anticipated/predicted by testing.

For more sensitive surfaces including the deteriorated aluminium honeycomb of the Service Module on the JSC vehicle, 15,000 psig was selected, using a rotating nozzle equipped with four – 0.008 inch jewels and used at an angle of incidence of 0°, 2 to 3 inches from the surface,[16] at a dwell time of approximately 0.5 sec/in^2.

Figure 12 Detail image of coating removal utilizing waterjet on top of the S-1C stage of the JSC Saturn V.

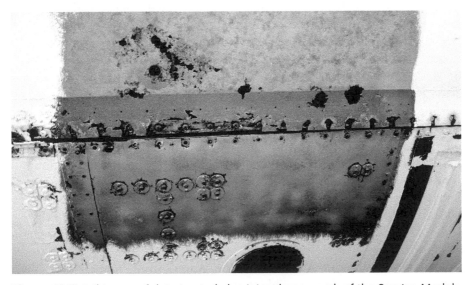

Figure 13 Detail image of deteriorated aluminium honeycomb of the Service Module after coating removal utilizing waterjets at 15,000 psig.

On all other areas of the rockets it was necessary to selectively vary pressures between 10,000 psig and 35,000 psig due to variations in the aluminium surface, profile, and thickness. Standards in stand-off distance were observed, except in heavily deteriorated areas of skin to reduce the risk of 'blow-through'.

Conclusion

The 'Big Piece' was previously treated with medium pressure water and the results were considered a success. This process was a departure from the labour intensive process of hand-picking with dental picks and scalpels, which was standard conservation practice at the time. The need for retreatment within a four year period was not prompted by a flaw in the treatment methods or materials used but rather from the improper protection of the artefact after treatment. The ultra-high pressure waterjet system was chosen because it provided the client with an artefact that was capable of withstanding the rigors of travel and less than ideal storage/display conditions while still retaining much of its desirable patina and sense of history.

The Saturn V rockets presented a unique challenge with their great size and complexity. They provided a unique opportunity that allowed the team to apply museum quality conservation standards to artefacts that had been previously treated by industrial painting contractors. The scale of the Saturn V rocket realistically eliminated the ability to treat the rocket by traditional conservation practices and previous treatment methods involving abrasive blasting techniques were immediately ruled out as being too aggressive. Ultra-high pressure waterjetting provided a large scale surface preparation option that would remove coatings, corrosion, and soluble salts while retaining the existing surface profile of the aluminium substrate. Accordingly, the project was able to meet both its budgetary constraints and strict environmental policies imposed upon the team by the respective clients.

While there are levels of quality and standards that we work towards in the conservation field, financial limitations and budget constraints are undeniable constants in our business. Most organizations do not have the financial resources to fully treat or even maintain their artefacts annually for an indefinite period of time, so it is our goal as conservators to continue to strive towards treatments and technologies that allow us to work more efficiently while still upholding the highest standards of practice. On these projects, ultra-high pressure waterjetting provided the means to achieve this goal.

The collaboration of members of the waterjetting industry and the conservation field made it possible to investigate the use of high and ultra-high pressure waterjets on historic industrial artefacts. This progress has potential benefits in many other areas of conservation, in particular out-door sculpture and architectural artefacts. However, additional testing and experimentation is always warranted prior to applying any developing technology to the field of conservation.

Acknowledgements

The authors would like to thank the following individuals for their contributions to this paper: Dawn Martin, Marvin Boatman (Boatman Industries, Inc., Houston, TX), Rick Watson (Surmac, San Antonio, TX) and the staff of R.M.S. *Titanic*, Inc.

Biography

Joseph Sembrat, President and Senior Conservator, Professional Associate, AIC
Joe Sembrat has been immersed in the conservation field for over 14 years providing conservation assessments, design, and implementation of conservation treatments and lecturing on relevant topics in the field. His extensive experience in the treatment of masonry and metals led to the development and creation of the unique firm Conservation Solutions, Inc (CSI) which he co-founded with his wife, Julya, in 1999 excelling in the treatment of historic monuments and sculpture, industrial artefacts, and buildings. High-profile projects include artefacts from the salvaged R.M.S. *Titanic* wreck-site, such as the 'Big Piece', the conservation of two Saturn V rockets, the treat-ment of 12 sets of over life-sized bronze gates at the US Commerce Department Building in Washington, DC, and a nineteenth-century Cotton Gin facility located in Scott, Arkansas. Sembrat achieved his Professional Associate status in AIC in 1996 and served as the Architectural Specialty Group Program Coordinator and Chair from 2000 to 2002.

Patty Miller, Conservator, Professional Associate, AIC
Patty Miller has been an active participant in the preservation field for more than ten years. As a conservator and project manager for CSI, Patty directs surveys, investiga-tions, testing programs and treatments of historic architecture, monuments and arte-facts. Since joining Conservation Solutions Inc. in 2003, she has been pivotal in the expansion of the in-house testing laboratory which enables CSI to service such diverse projects as the restoration of two Saturn V Rockets, the conservation of hundreds of artefacts salvaged from the R.M.S. *Titanic* wreck site, including the 17-ton 'Big Piece' of the *Titanic* hull, and numerous analyses of historic building materials dating from the seventeenth to twentieth century. Patty Miller's breadth of experience in conserva-tion and her dedication to education within the field of preservation is evident by her resume of lectures and papers. Most recently, Patty has had the opportunity to partici-pate as an instructor at the 2004 Cemetery Monument Conservation Seminar and Workshop held by the National Center for Preservation Technology and Training, and co-author of a paper on 'The Use and Effectiveness of Dispersed Hydrated Lime in the Conservation of Monuments and Historic Structures'.

Lydia M. Frenzel, Chief spokesperson for the Advisory Council
Lydia M. Frenzel is a professional member of the National Speakers Association. Lydia is a native Texan with a Ph.D. in Chemistry from the University of Texas, 1971. Lydia

donates significant time, primarily as a Chairperson, to standards organizations such as ISO, NACE, and SSPC, and is a director for 1995–2007 for the WaterJet Technology Association. Among her recognitions are: Steel Structures Painting Council 1996 Technical Achievement Award; Alpha Gamma Delta Sorority, Distinguished Citizen Award, 1997, and Rotary International District 5190 Governor 1997–98, and selected by JPCL as one of the 'Twenty people who have helped change the protective and marine coatings industry in the past 20 years'.

Jee Skavdahl, Project Manager
Jee Skavdahl joined Conservation Solutions, Inc. in December 2003 as a Conservation Technician and now serves the firm as a Project Manager for the Saturn V rocket conservation project at the Johnson Space Center in Houston, Texas. She has been closely involved in the management of both the treatment of the Saturn V and in the construction of the temperature and humidity-controlled building which currently houses and protects the Saturn V during the preservation process. Jee Skavdahl holds a Bachelor of Science in Physiological Science from the University of California, Los Angeles, California.

Notes

1   The aircraft industry, i.e. Delta, Lufthansa, KLM, United, American, Northwest, routinely use 36,000 psig to remove coatings while retaining the original profile on aluminium without distortion. One example is Pratt and Whitney 'Advanced Robotics Maintenance Systems (ARMS TM)' which operates at 28,000 psig. In working with General Electric Aircraft Engineers on the cleaning of engine components, Flow International calculated that it takes 158 times more energy to erode metal than is commonly used in surface cleaning (1995). NASA testing is documented in Spinoff 1996, NASA Center for AeroSpace Information (CASI) (ed.) 'Robotic WaterJet Systems', available at www.sti.nasa.gov. This article describes how, in the early 1990s, similar waterjet technology was developed when NASA teamed with United Technologies to produce an automated robotic maintenance system (ARMS) that uses 28,000 psig to strip thermal protection materials, paints and primers, layer by layer.

2   In 1999, R. K. Miller and G.J. Swenson of Thiokol presented 'Erosion of Steel Substrates when Exposed to Ultra-Pressure Waterjet Cleaning Systems' (WJTA, 10th American Waterjet Conference, August, 1999, paper 52, p. 661). Thiokol uses waterjetting to clean critical rocket engine parts for outer space. Prior to using UHP WJ (waterjetting), they had been cleaning the surfaces pits with dental picks. UHP WJ cleans these pits. Thiokol was concerned with damage to the metal substrate. The experiment used a target of D6AC steel with different sweep rates and rotation rates (dwell time). The plate was weighed before and after the sweep and an average profile was calculated. Data had been collected in 1992 and 1995 at 36,000 psig, and in 1998 with 40,000 psig, 1,300 revolutions per minute, standoff of 2.5 inch nominal, sweep rate of 60 inches per minute nominal. Thiokol verified that the minimum allowable erosion of 0.0001 inch (2.5 micron) would not be exceeded during the cleaning process. A grit blast of zirconium silicate produced an average profile of 0.700 mil (18 micron). A single pass of 40,000 psig produced an average profile of 0.009 mil; a second pass produced 0.017 mil; three to six passes produced an average profile of 0.018 mil (0.5 micron). Thus for 40,000

psig, this paper established that two passes eroded whatever material was going to be eroded from the D6AC steel. Then the steel substrate remained constant. These results are different from the 1992 tests. Thus, UHP WJ does make a micro-profile, but not of the same magnitude as the abrasive.

3   Surface profile, roughness or anchor pattern for coatings is the maximum peak-to-valley depth created during the cleaning, blasting, and preparation stages. This term is commonly associated with abrasive blast cleaning and is the result of the impact of the abrasive on the substrate.

4   Historical reference: Bryan, E.L. 'High Energy Jets as a new concept for wood machining,' Forest Products Journal, Vol 13, No. 8, August, 1963, p. 305.

5   PSI (pressure measured in pounds per square inch) is referred to in the waterjetting industry more specifically as PSIG, the pressure as it is measured at the pump gauge (pounds per square inch at gauge) where the operator controls are located. By following system parameters, the gauge pressure directly relates to the stream pressure impacting the work surface. The conversion for PSI to BAR and MPa is as follows: 1,000 psig = 68.9 BAR = 6.9 Mpa.

6   Merk-Gould, L.; Herskovitz, R. and Wilson, C., 'Field Tests on Removing Corrosion from Outdoor Bronze Sculptures Using Medium Pressure Water', ICOM, Vol. II, 1993, pp. 772–8.

7   Draughon, R., 'Ultrahigh-Pressure Blasting', NACE Corrosion 89, paper 119, 28 pp.

8   David Summers, Waterjetting Technology Chapman & Hall, New York, NY (1995) is a comprehensive treatise on civil engineering applications with chapters on how abrasives and water affect surfaces.

9   The unit of measure for the North American Waterjet Industry equipment is English (inches). Metric conversions are approximated as 0.002 inch = 0.01 millimetre, 2 to 3 inches = 5 to 7.5 centimetres, and 1 sec/in^2 approximately 2 sec/5 cm^2.

10   www.fertan.co.uk accessed 21 September 2005.

11   www.cortecvci.com accessed 21 September 2005.

12   The coupon samples were small disks of aluminum taken from the S-IVB stage of the JSC Saturn V Rocket. Samples were sent for EDS (energy dispersive X-ray spectroscopy) analysis. All 12 disks were probed at 20 kV. Acquisition time per probe was 120 seconds at a working distance of 39mm. Each disk was given an overall probe at 10X, and (3) additional probes were taken at 5000X (thus test 1, 2, 3). The instrument used was a JEOL JXA-840A Scanning Electron Microscope (SEM) with a ThermoNORAN TN-5502 energy dispersive analytical attachment and NORAN Vantage spectrum processing software. The system is also fitted with a low atomic number detector (Pioneer Premium detector with a Norvar window). The elements in the columns are what were found on the surface after the cleaning test – quantitative results are not given in this table.

13   The unit of measure for the North American Waterjet Industry equipment is English (inches). Metric conversions are approximated as 0.001 inch = 0.03 millimetre, 2 to 3 inches = 5 to 7.5 centimetres, and 1 sec/in^2 approximately 2 sec/5 cm^2.

14   Joint Surface Preparation Standard, NACE No. 5/ SSPC-SP12 'Surface Preparation and Cleaning of Steel and Other Hard Materials by High and Ultrahigh-Pressure WaterJetting' (revised 2002).

15   Soft abrasives in a stream of pressurized water were also tested to remove paint coatings. Universal Minerals, Inc. of Houston, Texas, provided the abrasive materials as well as an abrasive induction system operating with pressurized water from 3,000 to 20,000 psig and their RIPP 3000 nozzle. The RIPP 3000 nozzle uses the Venturi effect to introduce their Maxxstrip abrasives into the WJ stream. Maxxstrip is a sodium bicarbonate in fine, medium, and coarse grades. Maxxstrip has a neutral pH and was used in the demonstration.

16   The unit of measure for the North American Waterjet Industry equipment is English (inches). Metric conversions are approximated as 0.008 inch = 0.2032 millimetre, 2 to 3 inches = 5 to 7.5 centimeters, 0.5 sec/in^2 approximately 1 sec/ 5 cm^2

JOURNAL OF
Architectural Conservation

The international journal for historic buildings, monuments and places

Patron: Sir Bernard Feilden

Consultant Editors:
Professor Vincent Shacklock
Elizabeth Hirst
Professor Norman R. Weiss
Bob Kindred MBE

The scope of this international journal is intended to be wide-ranging and include discussion on aesthetics and philosophies; historical influences; project evaluation and control; repair techniques; materials; reuse of buildings; legal issues; inspection, recording and monitoring; management and interpretation; and historic parks and gardens.

© Donhead Publishing 2005

Editorial, Publishing and Subscriptions
Donhead Publishing Ltd
Lower Coombe, Donhead St Mary
Shaftesbury, Dorset SP7 9LY, UK
Tel: +44 (0)1747 828422
Fax: +44 (0)1747 828522
E-mail: jac@donhead.com
www.donhead.com

Managing Editor: Jill Pearce
Publishing Manager: Dorothy Newberry

Journal of Architectural Conservation is published three times per year.
Annual institutional subscription 2005: £94.00
Annual personal subscription 2005: £47.00
Back volumes and single issues are available from Donhead Publishing.

Papers appearing in the *Journal of Architectural Conservation* are indexed or abstracted in *Art and Archaeology Technical Abstracts, Avery Index to Architectural Periodicals, British Humanities Index, Art Index* and *Getty Conservation Institute Project Bibliographies.* Abstracts to all papers can be viewed on the Donhead website: **www.donhead.com**

ISSN: 1355-6207

J O U R N A L   O F
Architectural Conservation
The international journal for historic buildings, monuments and places

Notes for Contributors

PREPARATION AND SUBMISSION OF PAPERS

Authors should submit two paper copies and one electronic copy of the paper including illustrations, and should retain an additional copy for their own reference. Contributions for publication in the *Journal of Architectural Conservation* should be addressed to:

The Managing Editor
Journal of Architectural Conservation
Donhead Publishing Ltd
Lower Coombe, Donhead St Mary
Shaftesbury, Dorset SP7 9LY, UK

Tel: +44 (0)1747 828422
Fax: +44 (0)1747 828522
E-mail: jac@donhead.com

Submission of a paper to the Journal is taken to imply that it represents original work, which is not under consideration for publication elsewhere and has not been published previously. Papers and contributions published become the legal copyright of the publisher, unless otherwise agreed. All papers will be critically assessed by the team of Consultant Editors, and then peer reviewed before being accepted for publication.

Papers should be written in English and be between 2,000 and 5,000 words in total length (including abstract, biography and notes), double spaced and printed on one side of the paper (A4/letter). Shorter/viewpoint papers can be of 750 – 2000 words in length. Please note that typists should leave only one space after a full stop. Do not use automatic features such as end notes or comments. Our preferred format is on CD as a PC compatible file in Word version 2000, or earlier. If in doubt please check with the Publishing Manager, Dorothy Newberry.

The first page should contain: the title of the paper; the author's name, qualifications and affiliation; a brief biography (30–75 words); a descriptive abstract (100–150 words); and up to five keywords. Please provide full postal and e-mail addresses, telephone and fax numbers for all authors of the paper. Where a paper has in excess of three authors, the publisher may shorten the list with the use of *et al.*

Headings

Headings should not be numbered, and should be capitalized or marked (heading). Subheadings should be in bold or marked (subheading).

Illustrations

Up to ten illustrations per paper can be accepted, although if the paper requires more illustrations this can be discussed with the publishers. All illustrations must be suitable for black and white reproduction. Illustrations should accompany the typescript separately (i.e. not embedded in the text file). Figures and line drawings should be numbered consecutively (Fig 1, Fig 2 etc.), and submitted in their original form (not photocopies) ready for reproduction. Figure captions should be included within the main text of the paper, together with any credits. If digital images are to be supplied please contact Dorothy Newberry for our guidelines before submitting a sample illustration.

Photographs should be high quality positives, printed from original negatives, and suitable for black and white printing. The top of the illustration should be marked lightly in pencil on the back. Computer printouts of electronic images cannot be accepted.

Digital files of photographs need to be 300dpi at the final reproduction size and saved as TIF or JPEG files with minimal compression. The recommended format for digital line drawings is as bitmap files with a resolution of at least 600dpi at the final reproduction size. Tables and graphs should be submitted as an electronic file (Word or Excel) with a hard copy for reference only. The legend (key) and lines or shading must be suitable for black and white printing. If graphs are submitted as JPEG or TIF files they must be in bitmap format with a resolution of at least 600dpi at the final reproduction size.

References

References should be numbered consecutively through the text, then cited in full at the end of the paper in the following style:

for journals:
1. Jones, D.F. and Smith, B., 'Gauged Brickwork', *Journal of Brick Building*, Vol 4, No 1, June 1993, pp. 136–157.

for books:
2. Robson, P., *Structural Repair of Traditional Buildings*, Donhead, Shaftesbury (1999).

If a reference to a website is included, please indicate the date that it was accessed.

References should be typed in standard text format with numbers typed manually within the text, not produced using the automatic end notes function.

Milton Keynes UK
Ingram Content Group UK Ltd.
UKHW051924141024
449569UK00027B/1341